Reliability Analysis Using MINITAB and Python

Reliability Analysis Using MINITAB and Python

Jaejin Hwang
Northwestern University, USA

This edition first published 2023
© 2023 John Wiley & Sons, Inc. All rights reserved.

Published by John Wiley & Sons, Inc., Hoboken, New Jersey.

No part of this publication may be reproduced, stored in a retrieval system, or transmitted in any form or by any means, electronic, mechanical, photocopying, recording, scanning, or otherwise, except as permitted under Section 107 or 108 of the 1976 United States Copyright Act, without either the prior written permission of the Publisher, or authorization through payment of the appropriate per-copy fee to the Copyright Clearance Center, Inc., 222 Rosewood Drive, Danvers, MA 01923, (978) 750–8400, fax (978) 750–4470, or on the web at www.copyright.com. Requests to the Publisher for permission should be addressed to the Permissions Department, John Wiley & Sons, Inc., 111 River Street, Hoboken, NJ 07030, (201) 748–6011, fax (201) 748–6008, or online at https://www.wiley.com/go/permission.

Trademarks: Wiley and the Wiley logo are trademarks or registered trademarks of John Wiley & Sons, Inc. and/or its affiliates in the United States and other countries and may not be used without written permission. All other trademarks are the property of their respective owners. John Wiley & Sons, Inc. is not associated with any product or vendor mentioned in this book.

Limit of Liability/Disclaimer of Warranty: While the publisher and author have used their best efforts in preparing this book, they make no representations or warranties with respect to the accuracy or completeness of the contents of this book and specifically disclaim any implied warranties of merchantability or fitness for a particular purpose. No warranty may be created or extended by sales representatives or written sales materials. The advice and strategies contained herein may not be suitable for your situation. You should consult with a professional where appropriate. Neither the publisher nor author shall be liable for any loss of profit or any other commercial damages, including but not limited to special, incidental, consequential, or other damages. Further, readers should be aware that websites listed in this work may have changed or disappeared between when this work was written and when it is read. Neither the publisher nor authors shall be liable for any loss of profit or any other commercial damages, including but not limited to special, incidental, consequential, or other damages.

For general information on our other products and services or for technical support, please contact our Customer Care Department within the United States at (800) 762–2974, outside the United States at (317) 572–3993 or fax (317) 572–4002.

Wiley also publishes its books in a variety of electronic formats. Some content that appears in print may not be available in electronic formats. For more information about Wiley products, visit our web site at www.wiley.com.

Library of Congress Cataloging-in-Publication Data
Names: Hwang, Jaejin, author.
Title: Reliability analysis using MINITAB and Python / Jaejin Hwang.
Description: Hoboken, New Jersey : John Wiley & Sons, 2023. | Includes bibliographical references and index.
Identifiers: LCCN 2022029036 (print) | LCCN 2022029037 (ebook) | ISBN 9781119870760 (hardback) | ISBN 9781119870777 (pdf) | ISBN 9781119870784 (epub)
Subjects: LCSH: Minitab. | Reliability (Engineering) | Python (Computer program language)
Classification: LCC TA169 .H93 2023 (print) | LCC TA169 (ebook) | DDC 620/.00452--dc23/eng/20220906
LC record available at https://lccn.loc.gov/2022029036
LC ebook record available at https://lccn.loc.gov/2022029037

Cover image: © whiteMocca/Shutterstock
Cover design by Wiley

Set in 9.5/12.5pt STIXTwoText by Integra Software Services Pvt. Ltd, Pondicherry, India

Contents

About the Author

Jaejin Hwang is an associate professor of industrial and systems engineering at Northern Illinois University. He has been teaching reliability engineering and advanced quality control courses since 2016. He has actively used various software including Minitab, Excel, Python, SPSS, and Matlab to promote students' learning. Dr Hwang holds a PhD in industrial engineering from Ohio State University. His research interests span the areas of quality and reliability, work measurement and work design, ergonomics, and occupational biomechanics. In 2022 he was nominated for the excellence in undergraduate teaching award at Northern Illinois University. He has authored over 50 technical papers published in peer-reviewed journals, international conference proceedings, and magazines. His book *Data Analytics and Visualization in Quality Analysis Using Tableau* was published by CRC Press in July 2021. Dr Hwang has been involved in numerous student graduation research and industrial projects. He is an executive committee member of the International Society for Occupational Ergonomics and Safety. He is an editorial board member of *Work: A Journal of Prevention, Assessment & Rehabilitation* and the Korean Society for Emotion and Sensibility. He is a guest editor for the special issue (November 2020–November 2022) in the *International Journal of Environmental Research and Public Health.*

Preface

Overview

Reliability is a vital and effective tool to analyze how long products and services can show satisfactory performance without failure. Today, we live in a society that uses complex and sophisticated physical and digital products. With the development of these technologies, the importance of the field of reliability is increasing.

This book is based on the statistical theory of how to analyze reliability. We want to quantitatively and accurately predict the reliability of products and services using various statistical distributions and probabilities. Reliability is the tool that helps to perform these analysis methods efficiently. Reliability tools allow us to process massive amounts of data quickly and automate our analysis methods efficiently. This book introduces you to how to perform reliability analysis using Minitab and Python.

Audience

- Students and professionals interested in the field of reliability
- Undergraduate or graduate students majoring in industrial engineering, mechanical engineering, electronic engineering, or related fields

Chapter Outline

Chapter 1, Introduction, introduces the basic concept and history of reliability. The causes for failure are introduced, and the reliability bathtub curve is explained.

Chapter 2, Basic Concepts of Probability, introduces the basic theory of probability. Probability calculation is performed through reliability-related examples.

Chapter 3, Lifetime Distribution, deals with statistical distributions frequently used in the field of reliability. Practice includes making graphs and flexibly modifying parameter values of statistical distributions using Minitab and Python.

In Chapter 4, Reliability Data Plotting, we practice finding appropriate statistical distributions by plotting reliability data and finding relevant parameter values. We learn to perform these procedures efficiently with Minitab and Python.

In Chapter 5, Accelerated Life Testing, we study the theory of accelerated life testing and learn to predict failure characteristics in an actual use environment through the accelerated life testing data. We learn to plot and make predictions about various stressful environments using the reliability tool.

Chapter 6, System Failure Modeling, calculates how the failure probability and distribution of various factors affect the failure characteristics of the entire system based on statistical methods.

Chapter 7, Repairable Systems, deals with preventive and corrective maintenance. It covers calculations to establish a schedule that minimizes costs in preventive maintenance.

In Chapter 8, Case Studies, reliability analysis is performed through various examples using Minitab and Python. Warranty analysis, non-parametric analysis, and stress–strength interference analysis that have not been covered in previous chapters are introduced.

Text Material

Minitab, Python, and Excel files related to the examples used in this book are provided. PowerPoint lecture materials for lecturers are also provided.

Acknowledgments

I would like to thank the staff of John Wiley, who readily accepted the proposal for this book. In writing *Reliability Analysis*, I was inspired by the writings and resources of many reliability experts. I would like to express my gratitude to the following people: the authors of the book *Applied Reliability* (Paul A. Tobias and David C. Trindade) and the developers of the Python library related to reliability (Matthew Reid and Derryn Knife). This book is based on materials from reliability engineering classes I have taught over five years. I would like to thank all Northern Illinois University students who took my courses for providing valuable feedback to ensure that the course material continues to evolve.

About the Companion Website

This book is accompanied by a companion website which includes a number of resources created by author that you will find helpful.

www.wiley.com\go\Hwang\ReliabilityAnalysisUsingMinitabandPython

The website includes the following resources for each chapter:

- Minitab
- Python
- Excel

1

Introduction

Chapter Overview and Learning Objectives

- To understand the importance of the field of reliability in modern society.
- To learn the history and motivation of the field of reliability.
- To understand the definitions of reliability and failure.
- To explore several causes of failures.
- To understand different types of failure time.
- To learn the concept of the reliability bathtub curve.

1.1 Reliability Concepts

1.1.1 Reliability in Our Lives

In modern society, we rely on various complex and advanced devices and systems to enjoy convenience and enhance our lives. For example, autonomous vehicles allow drivers to engage in other activities while driving, and advances in space technology have ushered in an era in which civilians can also go to space.

However, advances in technology have some side effects. With the development of technology, the area in which humans can intervene is reduced, and when there is a problem in a system or device, humans can be adversely affected. For example, cars that are being produced these days are often electronic systems that systematically operate the overall functions of the car. If the central computer unit that controls the car breaks down, the driver will be limited in what they can do with it. Autonomous vehicle accidents due to system errors are also occasionally encountered, which can be viewed in a similar context.

With the development of these technologies, the necessity of the reliability field can be felt more acutely. A complex system may have thousands or tens of

Reliability Analysis Using MINITAB and Python, First Edition. Jaejin Hwang.
© 2023 John Wiley & Sons, Inc. Published 2023 by John Wiley & Sons, Inc.
Companion Website: www.wiley.com\go\Hwang\ReliabilityAnalysisUsingMinitabandPython

thousands of large and small parts interlocked. If the failure characteristics of such a system and the time it takes to failure can be predicted, very serious accidents can be prevented, and customer satisfaction can be increased.

1.1.2 History of Reliability

When did the field of reliability begin to gain attention? The United States is one of the major countries that pioneered the field of reliability. The need for reliability was highlighted during World War II. Electronic military equipment was shipped immediately after development, but when it arrived at the destination (far east), the equipment underwent many failures. At the time of development, there was no defect when the quality inspection was conducted. After investigation, it was found that the place where the equipment was actually used was a high-temperature and high-humidity environment, which caused equipment (vacuum tube) failure. In other words, the environment in which the equipment was developed and the environment in which it was used were very different, causing failure. At this time, a professional reliability analysis team was founded, and efforts to reduce the occurrence of failures in the actual equipment-use environment began.

At the same time, Germany also began to actively consider using the reliability field to increase the mission success rate of missiles. In modern society, reliability-related fields are widely used in addition to their use in the military industry. Reliability is considered to be an extended concept of quality control, and it is not an exaggeration to say that reliability is considered in designing and manufacturing almost all equipment and products, from industrial equipment to household appliances.

1.1.3 Definition of Reliability

According to the dictionary definition of the word reliability, it is as follows:

> The quality of being trustworthy or of performing consistently well.

This definition can also be considered in connection with product or system failure. It can be said that **minimizing failure** is one of the important factors of reliability.

In addition, **time** is another very important factor in defining reliability, because reliability is dependent on time. For example, the failure rate within 1 year of buying a car and after 10 years will be very different. For this reason, the concept of time is an essential element in reliability.

Finally, reliability can be expressed as a **probability**. That is, it can be evaluated through a quantified method. For example, the probability that the purchased car will fail within 1 year could be 5%, and the probability that it will fail within 10 years could be 90%.

In summary, reliability can be defined as:

> The probability that a system or product will perform the expected function in a specific environment over a specific period of time.

In other words, the defect in military equipment (vacuum tubes) in World War II previously mentioned can be seen as a result of the lack of consideration for the specific environment in which the equipment was used. In order to understand and predict reliability in such a case, the concept of probability and distribution of statistics can be applied, and it will be mainly covered in this book.

Case
In the US auto market, the warranty is usually 3 years or 36,000 miles. Automaker A is planning a 5-year or 60,000-mile warranty to give it an edge over its competitors. The automaker wants to know whether the parts they receive from their subcontractors are 99% or more reliable over 5 years. How can subcontractors prove the reliability of parts?

1.1.4 Quality and Reliability

Quality and reliability are often considered together, but the difference between these two concepts needs to be explored, as shown in Table 1.1.

Table 1.1 Description of quality and reliability.

Quality	Reliability
Before a product is shipped, check it to be sure there are no defects and the product is satisfactory.	Evaluate how satisfactorily the product or system operates throughout the entire lifespan.
Assess whether the characteristics of the product or component are within the specification and whether the process is within control.	Focus on how long a product or system can maintain satisfactory functioning with minimal failures.

The American rating company J.D. Power and Associates considers the quality and reliability of automobiles as important metrics, and automotive customers are also interested in these metrics. The new car Initial Quality Index (IQS) is an index that quantifies the number of problems per 100 new cars within 90 days. The Vehicle Dependability Study Index (VDS) quantifies the number of problems per 100 vehicles that occurred in the previous year among vehicles older than 3 years. We can see by the stated time spans that the concept of reliability is implied.

1.1.5 The Importance of Reliability

As products and systems become more sophisticated and complex, reliability becomes more and more critical. In the case of a smartphone, more than 500 parts are normally included. Failure of certain parts can lead to serious failures that make the entire smartphone unusable. In addition, the complexity of products and systems can expose them to more human error in the design and the development stage.

Failure of products and systems has a substantial impact on safety and cost. According to the factor of 10 rule, the more delayed the response to the reliability problem is, the more the cost of tenfold or greater increases as it moves to the next stage, as seen in Figure 1.1.

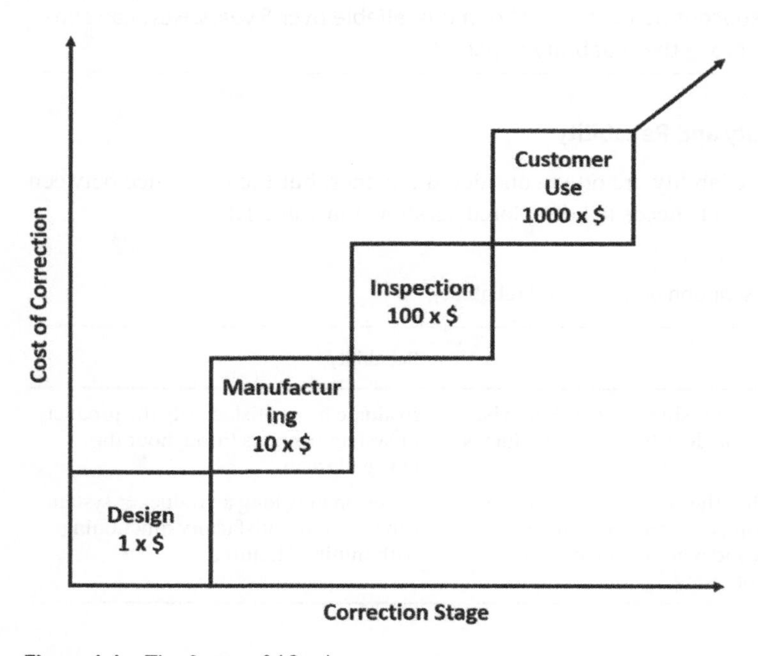

Figure 1.1 The factor of 10 rule.

The development cycle of products is gradually shortening due to advances in technology, heated competition for products, and increased customer expectations. In the case of smartphones, development cycles often range from six months to less than a year. Such a tight schedule increases the risk of human errors and failures. The importance of reliability concepts throughout the entire product cycle could grow in these circumstances.

1.2 Failure Concepts

1.2.1 Definition of Failure

The dictionary definition of failure is as follows:

Lack of success.
The omission of expected or required action.

If we approach the definition of failure from an engineering perspective, it is as follows:

The inability of a system or component to perform its required functions within specified performance requirements.

In other words, the concept of failure is closely related to the definition of reliability mentioned previously. Failure can also be classified into hard and soft failures, as seen in Table 1.2.

Table 1.2 Description of hard and soft failures.

Hard Failure	Soft Failure
Failure in which all system functions are lost due to sudden stop of function.	Failure that reaches the endurance limit due to gradual deterioration of operating characteristics and deterioration of performance.

Adverse effects caused by failure can be divided into three stages according to the scale shown in Figure 1.2. A small failure can cause inconvenience in our daily life, such as an air conditioning malfunction. Intermediate failures can take a toll on our bodies, property, safety, and finances, such as with a vehicle's brake failure. In addition, large failures cause disasters in society. Examples include breakdowns or crashes of airplanes and spacecraft.

1.2.2 Causes of Failure

There are many different causes of failure. Let's take a look at each of these causes.

1) Overstress
   If the external load is higher than the strength that the component can withstand, failure could occur. This could include mechanical overstress as well as electrical overstress. For example, the maximum load that an elevator can

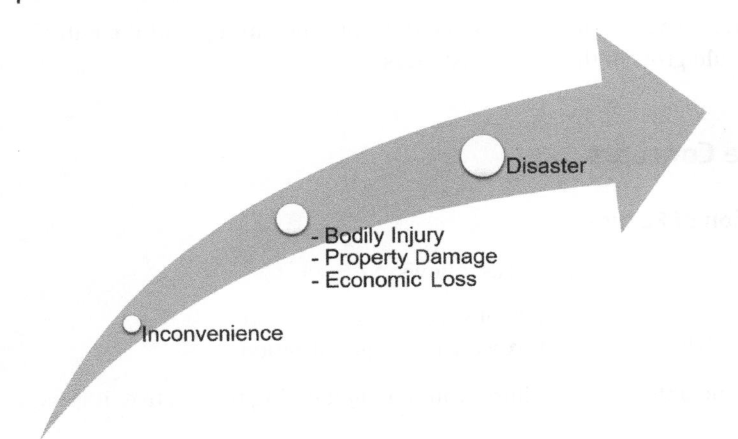

Figure 1.2 The adverse effects of failure.

withstand is specified. If a load higher than this maximum capacity is applied, a failure could occur in the elevator's mechanical system. Thus, an alarm system is activated when the load is exceeded to prevent failure.

2) Variation
The greater the variation in external loads and component strength, the greater the chance that failure could occur. Figure 1.3 shows the distribution of load and strength with no variation. The external load occurs consistently only at a specific value, and the strength of the component also continuously holds only a particular value. In this case, theoretically, the probability of a failure can be considered negligible.

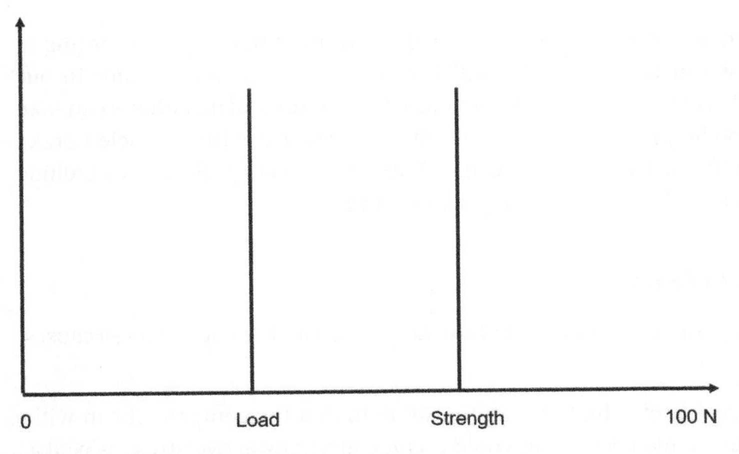

Figure 1.3 Frequency distribution of load and strength with no variation.

Figure 1.4 shows the distribution of the load and strength with considerable variation. There is an overlap between the load and strength distributions. In other words, the excessive load portion could be greater than the weakest strength portion, which could lead to failure. This indicates that a greater variation between the load and strength could increase the chance of failure.

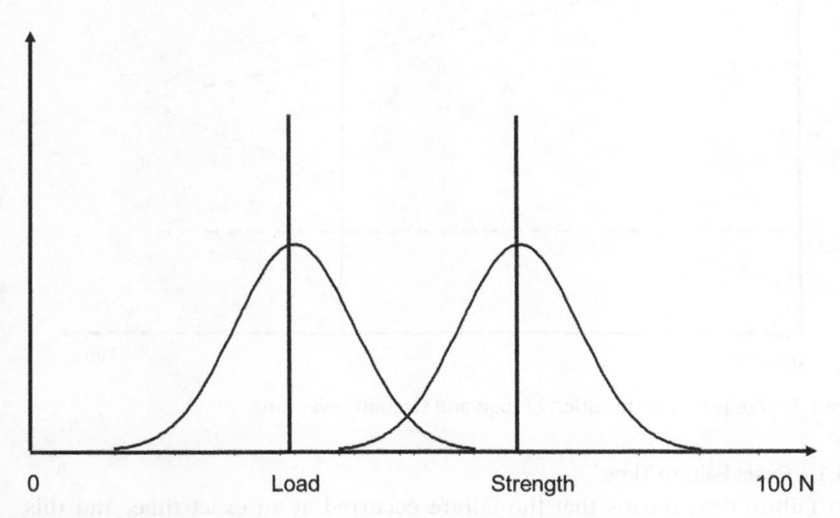

Figure 1.4 Frequency distribution of load and strength with variation.

3) Wearout
   Failure can also be affected by wearout or fatigue, which means that the load and strength are time-dependent. The magnitude and variations of the load and strength could change over time, which would cause the chance of failure. Figure 1.5 shows an example of wearout failure. The load's magnitude and variation are consistent over time. However, the strength's magnitude decreases, and the variation increases over time. After a certain period of time, the overlap between the load and strength can be observed, which indicates the increased chance of failure due to wearout.

1.2.3 Types of Failure Time

Failure time can be categorized into different types depending on the test methods and conditions.

- Exact failure time
- Right-censored failure time
- Left-censored failure time
- Interval-censored failure time

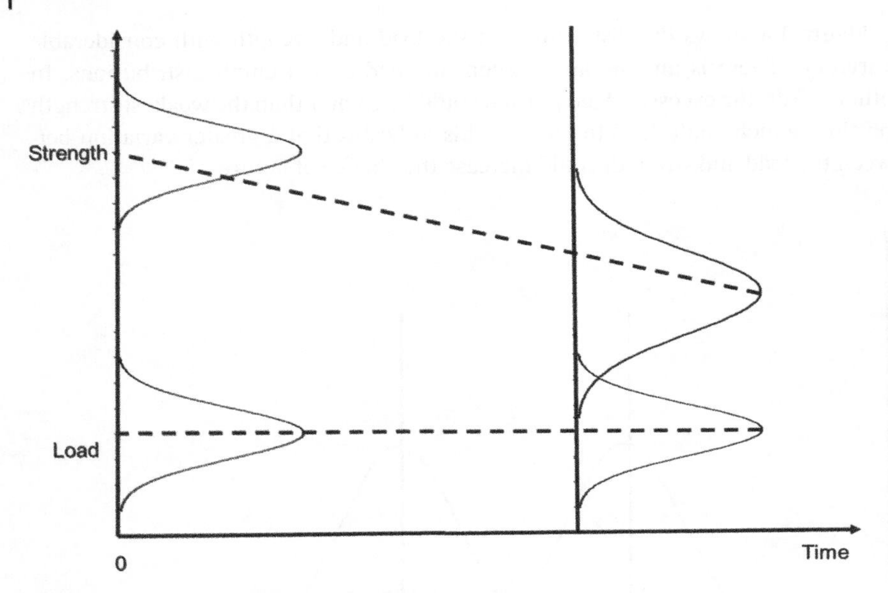

Figure 1.5 Frequency distribution of load and strength over time.

1.2.3.1 Exact Failure Time

Exact failure time means that the failure occurred at an exact time, and this failure time is observed and recorded. Figure 1.6 illustrates the exact failure time of three different parts. The advantage of the exact failure time is to obtain accurate information of when the components fail. The limitation of the exact failure time is time-consuming and often requires the cost of setting up measurement devices.

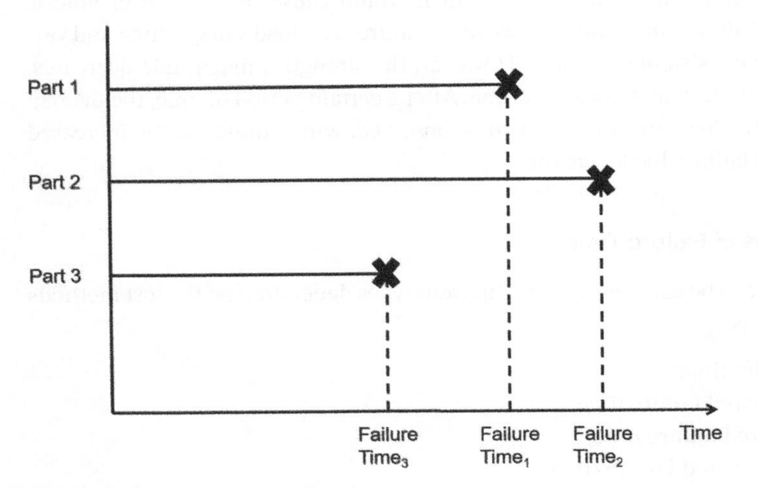

Figure 1.6 Description of the exact failure time.

1.2.3.2 Censored Failure Time

The censored failure time indicates that the exact failure time of the components is unknown or not observed during the test. Given the nature of highly reliable components, it is realistic and possible not to observe the failure of components during the test period. In this case, we denote this condition as a censored failure. Depending on the test methods and situations, there are three different censored failure times: right-censored failure time, left-censored failure time, and interval-censored failure time.

1.2.3.3 Right-Censored Failure Time

Right-censored failure time means that the component does not fail even after the completion of the test period. Figure 1.7 describes an example of right-censored failure time. At the time point the test period ends, part 2 has not failed and is still running. In this case, part 2's exact failure time is unknown. We use the test period's completion time as a right-censored failure time of part 2.

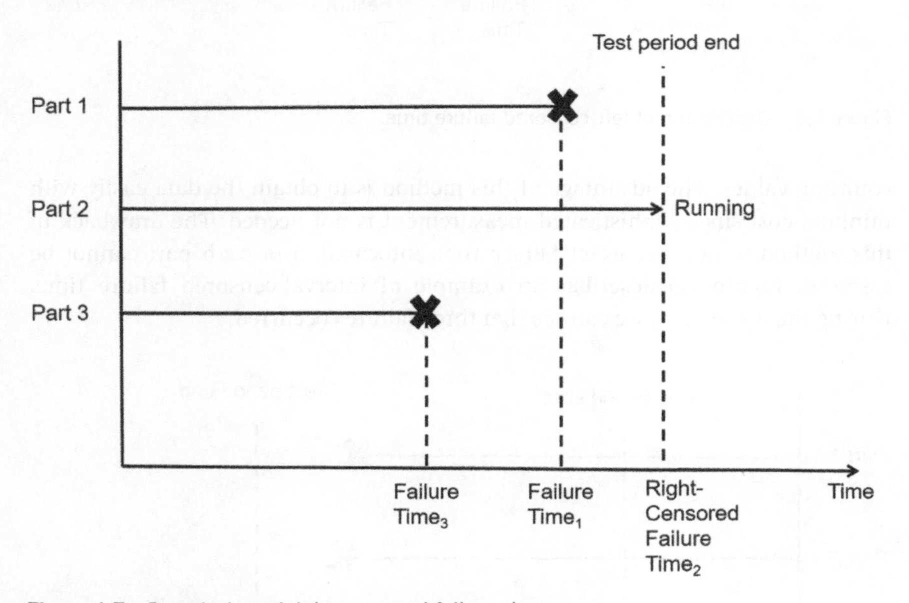

Figure 1.7 Description of right-censored failure time.

1.2.3.4 Left-Censored Failure Time

Left-censored failure time means that when the test period begins, some parts have already failed, so their exact failure time is unknown. Figure 1.8 illustrates left-censored failure time. When the test period begins, it is observed that part 2 has already failed. In this case, we do not know exactly when part 2 failed.

1.2.3.5 Interval-Censored Failure Time

Interval-censored failure time includes the number of failures during the test period. Unlike the previous failure time categories, this failure time deals with the

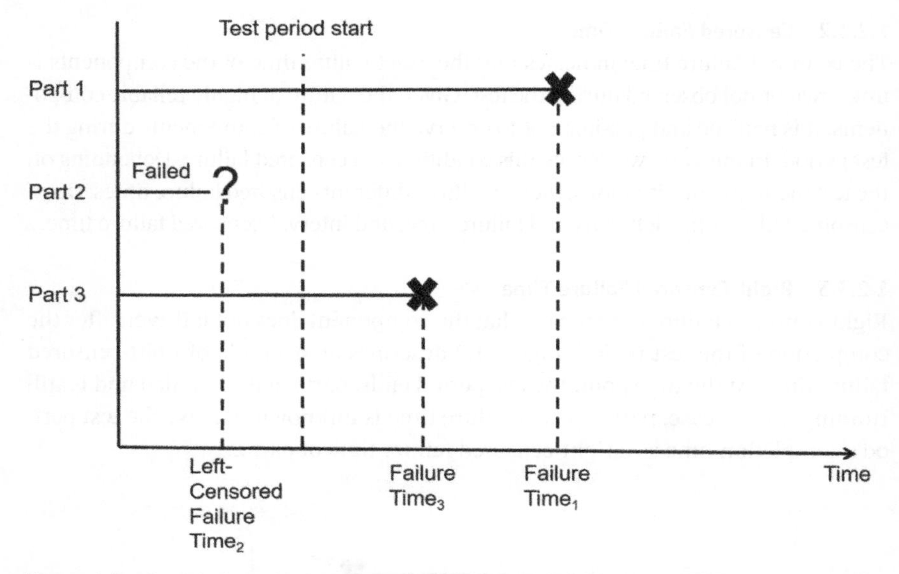

Figure 1.8 Description of left-censored failure time.

counting values. The advantage of this method is to obtain the data easily with minimal cost since sophisticated measurement is not needed. The drawback of this method is that the exact failure time information of each part cannot be assessed. Figure 1.9 describes an example of interval-censored failure time. During the test period, we can see that three failures occurred.

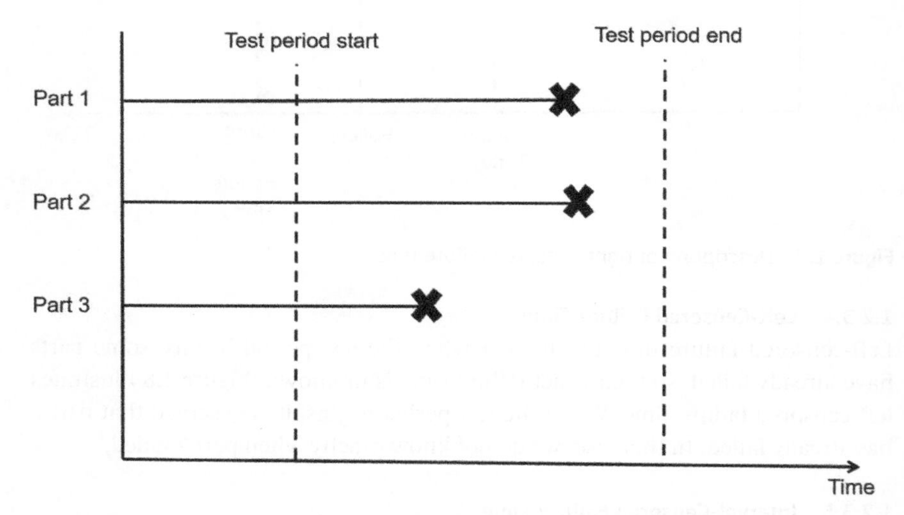

Figure 1.9 Description of interval-censored failure time.

1.2.3.6 Minitab Practice

With Minitab, we could set up the data by indicating the censored values. For example, there were 10 components tested to observe the failures. At the end of the testing period of 45 hours, 7 components failed, but 3 components still survived as seen in Table 1.3.

Table 1.3 Time to failure data of 10 components.

Component	Time to Failure (hours)	Censoring
1	5	No
2	8	No
3	15	No
4	32	No
5	38	No
6	42	No
7	43	No
8	Unknown	Yes
9	Unknown	Yes
10	Unknown	Yes

The data set can be prepared when using Minitab as seen in Figure 1.10. The censoring column is set, where 0 means non-censored value and 1 denotes the censored values. For the censored values, the end of the observation time (45 hours) can be input for the time to failure.

↓	C1	C2
	Time to failure (hours)	Censoring
1	5	0
2	8	0
3	15	0
4	32	0
5	38	0
6	42	0
7	43	0
8	45	1
9	45	1
10	45	1

Figure 1.10 Data set with Minitab.

1.2.3.7 Python Practice

Python can be used to manage the data with different types of failure time. Here is an example of treating right-censored failure times.

The Python codes can be written using the platform of Google Colab (https://colab.research.google.com). Here are the Python codes to manage the right-censored time data (Figure 1.11).

[pip install] allows installing the library resource of [reliability] (Reid, 2022).

[import make_right_censored_data] will help us to manage the right-censored failure time.

We could write the failure time in the data array. The threshold is the end of testing.

Although we do not know the actual failure time of the three survived components, we could put arbitrary values exceeding the threshold to treat them as censored values.

[print] allows us to display the uncensored and censored data.

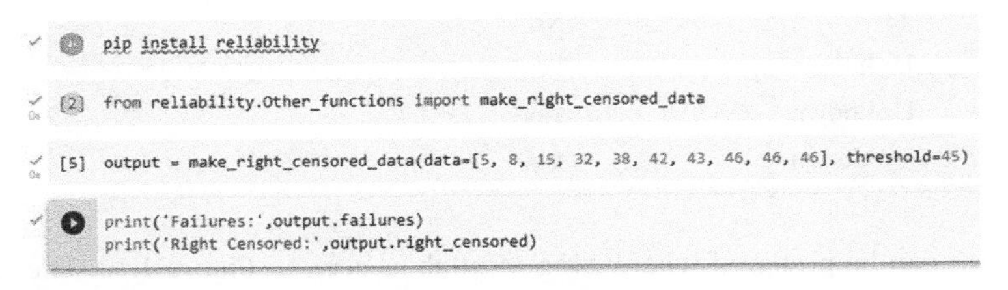

pip install reliability
[2] from reliability.Other_functions import make_right_censored_data
[5] output = make_right_censored_data(data=[5, 8, 15, 32, 38, 42, 43, 46, 46, 46], threshold=45)
print('Failures:',output.failures)
print('Right Censored:',output.right_censored)

Figure 1.11 Python codes to manage the right-censored failure time data.

After running all the codes, the results will be shown on the Google Colab. The last three components' right-censored failure time was assigned as 45 hours (Figure 1.12).

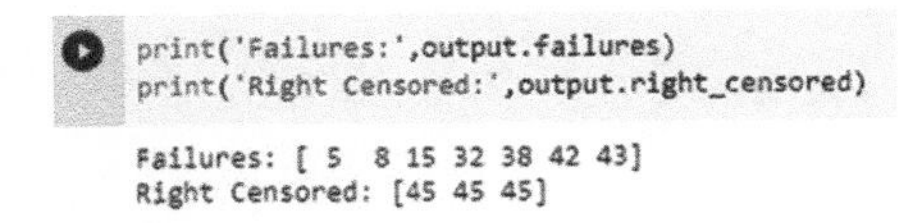

print('Failures:',output.failures)
print('Right Censored:',output.right_censored)
Failures: [5 8 15 32 38 42 43]
Right Censored: [45 45 45]

Figure 1.12 Right-censored data output using Python.

1.2.4 The Reliability Bathtub Curve

The reliability bathtub curve is used to represent the failure rate of the component over time. The failure rate curve looks like a contour of a bathtub, thus it is called a bathtub curve. The failure rate is also called a hazard rate. This measure indicates the number of failures per unit time (e.g., hour, day, week, month, year).

In the reliability bathtub curve, there are three different shapes of failure rate curves in different life stages, as seen in Figure 1.13.

- Early life
- Useful life
- Wearout life

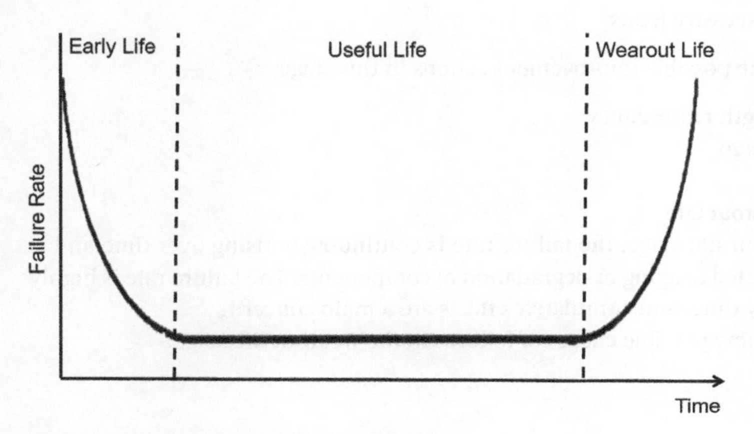

Figure 1.13 The reliability bathtub curve.

1.2.4.1 Early Life

In the early life stage, a high failure rate is expected at the beginning of product use. It is also called infant mortality. There is a rapidly decreasing pattern of the failure rate over time because defective items are investigated and continuously discarded at this stage. The burn-in tests in electronic parts can be performed in this early life stage. This stage would not be desirable to customers due to the high failure rate. It would substantially increase customer dissatisfaction and also increase warranty expense for the manufacturer.

Here are some possible causes of failures in the early life stage.

- Manufacturing defects
- Assembly errors
- Poor quality control
- Poor workmanship

Here are some possible improvement actions in this stage.

- Improved quality control
- Accelerated stress testing

1.2.4.2 Useful Life

In the useful life stage, a low and constant failure rate over time is expected. In other words, the failure rate is not affected by time. This stage could be ideal for shipping a product to customers. In this stage, random failures mainly occur. The design goal is to maintain the failure rate as low as possible throughout the time period.

Here are some possible causes of failures in the useful life stage.

- Environment
- Human errors
- Random excessive loads

Here are some possible improvement actions in this stage.

- High strength redundancy
- Robust design

1.2.4.3 Wearout Life

In the wearout life stage, the failure rate is continuously rising over time and is primarily related to aging or degradation of components. The failure rate is highly influenced by time, and cumulative effects are a main concern.

Here are some possible causes of failures in the wearout life stage.

- Fatigue
- Corrosion
- Aging
- Friction

Here are some possible improvement actions in this stage.

- Preventive maintenance
- Replacement
- Robust material

1.2.4.4 Python Practice

Python can be used to display the reliability bathtub curve. Google Colab (https://colab.research.google.com) can be utilized to run the Python codes.

[pip install reliability] would install the library resource of the reliability package.

[matlpotlib] can also be installed to generate various visualizations.

[reliability.Distributions] source can be used to generate several lifetime distributions. These distributions will be discussed later.

[matplotlib.pyplot] and **[numpy]** can be used to deal with the data in a specific array, and generate the plots.

[np.linspace] determines the start, end, and interval of the data range.

[Weibull_Distribution] with specific parameters can be used to construct the plot of the early stage (infant_mortality).

[Exponential_Distribution] with a specific parameter is used to construct the plot of the useful life stage (random_failures).

[Lognormal_Distribution] with specific parameters is used to construct the plot of the wearout life stage.

Figure 1.14 shows the Python codes used to construct the reliability bathtub curve.

```
pip install reliability

[ ] pip install matplotlib==3.1.3

[ ] from reliability.Distributions import Weibull_Distribution, Lognormal_Distribution, Exponential_Distribution

[ ] import matplotlib.pyplot as plt
    import numpy as np

[ ] xvals = np.linspace(0,1000,1000)

[ ] infant_mortality = Weibull_Distribution(alpha=400,beta=0.7).HF(xvals=xvals,label='infant mortality [Weibull]')
    random_failures = Exponential_Distribution(Lambda=0.001).HF(xvals=xvals,label='random failures [Exponential]')
    wear_out = Lognormal_Distribution(mu=6.8,sigma=0.1).HF(xvals=xvals,label='wear out [Lognormal]')
    combined = infant_mortality+random_failures+wear_out
    plt.plot(xvals,combined,linestyle='--',label='Combined hazard rate')
    plt.legend()
    plt.title('Bathtub curve')
    plt.xlim(0,1000)
    plt.ylim(bottom=0)
    plt.show()
```

Figure 1.14 Python codes used to construct the reliability bathtub curve.

After running all codes, the reliability bathtub curve can be created, as shown in Figure 1.15.

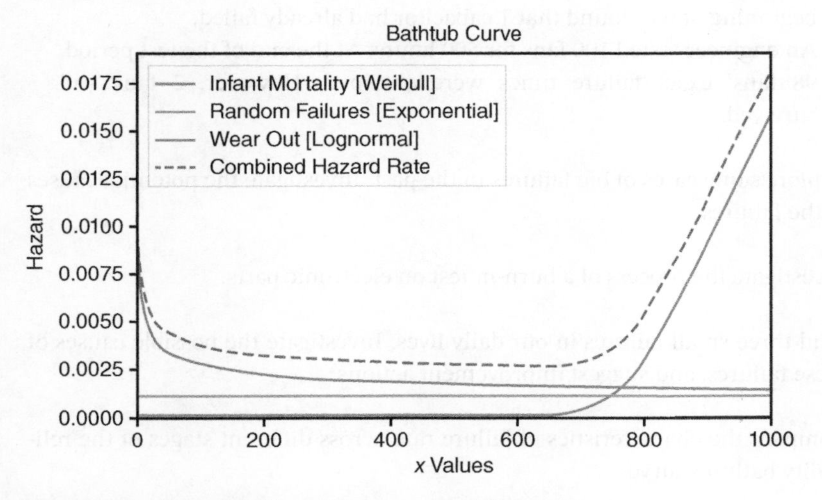

Figure 1.15 The reliability bathtub curve with Python.

1.3 Summary

- With the development of modern and complex technologies, the necessity of the reliability field can be felt more acutely.
- The need for reliability was highlighted during World War II.
- Reliability is the probability that a system or product will perform the expected function in a specific environment over a specific period of time.
- Failure is the inability of a system or component to perform its required functions within specified performance requirements.
- There are many different causes of failure, such as overstress, variation, and wearout.
- The failure time can be categorized into different types depending on the test methods and conditions, including exact failure time, right-censored failure time, left-censored failure time, and interval-censored failure time.
- In the reliability bathtub curve, there are three different shapes of failure rate curves in different life stages including early life, useful life, and wearout life.

Exercises

1 Determine the proper type of failure time for each of the following cases.

A An engineer inspects the components every 10 hours and records the number of failures in each interval.

B The reliability engineer started the inspection of 10 capacitors. In the beginning, it was found that 1 capacitor had already failed.

C An engineer tested 100 fans for 500 hours. At the end of the test period, 98 fans' exact failure times were obtained. However, 2 fans still survived.

2 Explore some cases of big failures in the past. Investigate the potential causes of the failures.

3 Investigate the process of a burn-in test on electronic parts.

4 Find three small failures in our daily lives. Investigate the possible causes of these failures, and suggest improvement actions.

5 Compare the characteristics of failure rate across different stages of the reliability bathtub curve.

Reference

Reid, M. (2022). Reliability—a Python library for reliability engineering (Version 0.8.1) [Computer software]. Zenodo. Available at: https://doi.org/10.5281/ZENODO.3938000 [Accessed 06/27/2022].

2

Basic Concepts of Probability

Chapter Overview and Learning Objectives

- To understand the fundamental concepts of probability.
- To understand the differences between various probability rules.
- To apply the probability rules to reliability examples.
- To learn the required assumptions of the probability rules.

2.1 Probability

Probability is a ratio of specific outcomes to total possible outcomes. In our daily lives, we hear from the news about the chance of rainfall, which could be an example of probability. The equation for probability is shown here.

$$P(\mathrm{A}) = \frac{n}{N} \tag{2.1}$$

The n indicates the number of elements in events. The N denotes the number of elements in the sample space. Thus, $P(\mathrm{A})$ means the probability of obtaining an outcome A. Probability is always ranged between 0 and 1 (Equation 2.2). If $P(\mathrm{A}) = 0$, the event cannot occur. If $P(\mathrm{A}) = 1$, a sure event is expected.

$$0 \leq P(\mathrm{A}) \leq 1 \tag{2.2}$$

Example 2.1 If you flip a coin, what are the possible outcomes? If you are interested in seeing a tail, what is the probability?

Answer:
Possible outcomes = {Head, Tail}.

Reliability Analysis Using MINITAB and Python, First Edition. Jaejin Hwang.
© 2023 John Wiley & Sons, Inc. Published 2023 by John Wiley & Sons, Inc.
Companion Website: www.wiley.com\go\Hwang\ReliabilityAnalysisUsingMinitabandPython

$$P(\text{Tail}) = \frac{1}{2}$$

2.1.1 The Importance of Probability in Reliability

The probability concept can be effectively applied to the field of reliability. Quantitative estimation of the chance of failure of components or systems could be useful information. Not only can we estimate the probability of a single component, but we can also use probability to estimate the entire system's reliability. Probability concepts such as joint probability and union probability can be used to understand the whole system's chance of failure or survival.

Example 2.2 What is the chance of seeing two defective components when you inspect three components?

Answer:
Let's denote that D = defective component, and G = good component.
The sample space is {DDD, DGD, DDG, GDD, DGG, GDG, GGD, GGG}.
Event A is getting exactly two defective components: {DGD, DDG, GDD}.

$$P(A) = \frac{3}{8}$$

2.2 Joint Probability with Independence

Joint probability is the probability of multiple events occurring simultaneously. In the Venn diagram, it can be an interaction of multiple events as seen in Figure 2.1. For two events, it can be written as $P(A \cap B)$ or $P(\text{A and B})$.

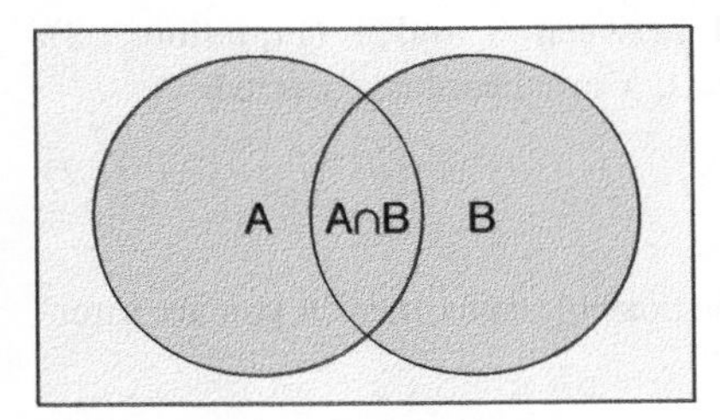

Figure 2.1 Joint probability.

If multiple events are independent of each other, the multiplication rule can be applied. The probability of each independent event can be simply multiplied together. Equation 2.3 shows the joint probability of two independent events.

$$P(\mathrm{AB}) = P(\mathrm{A})P(\mathrm{B}) \tag{2.3}$$

Example 2.3 In a manufacturing plant, there are two main machines. For machine A, the probability of producing a defective part is 0.03. For machine B, the probability of seeing a defective part is 0.05. The two machines are independently operating. What is the probability that both machines will produce defective parts simultaneously?

Answer:
Event A = a defective part from machine A
Event B = a defective part from machine B

$$P(\mathrm{AB}) = P(\mathrm{A})P(\mathrm{B}) = 0.03 \times 0.05 = 0.0015$$

2.3 Union Probability

Union probability is the probability that either of multiple events may occur. For two events, it can be written as $P(\mathrm{A} \cup \mathrm{B})$ or $P(\mathrm{A}$ or $\mathrm{B})$. In a Venn diagram, it can be a whole area of multiple events as seen in Figure 2.2.

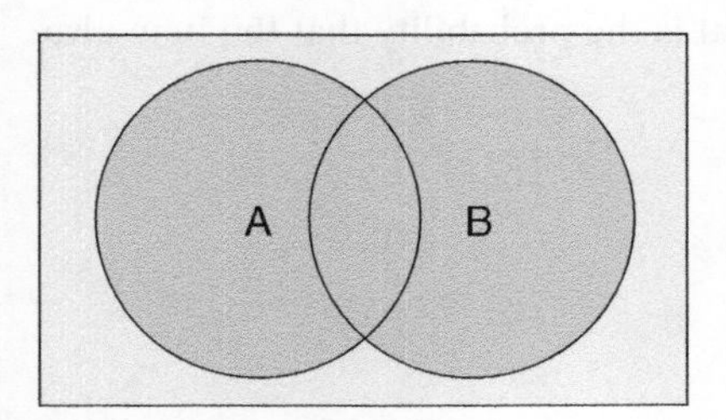

Figure 2.2 Union probability.

The equation of the union probability of two events is shown here.

$$P(\mathrm{A} \cup \mathrm{B}) = P(\mathrm{A}) + P(\mathrm{B}) - P(\mathrm{AB}) \tag{2.4}$$

Example 2.4 In a manufacturing plant, there are two main machines. For machine A, the probability of producing a defective part is 0.03. For machine B, the probability of seeing a defective part is 0.05. The probability that both machines produce defective parts simultaneously is 0.0015. What is the probability that machine A or machine B will produce a defective part?

Answer:
Event A = a defective part from machine A
Event B = a defective part from machine B

$$P(A \cup B) = P(A) + P(B) - P(AB) = 0.03 + 0.05 - 0.0015 = 0.0785$$

2.4 Conditional Probability

Conditional probability is the probability of an event given that another event has occurred. For example, $P(A|B)$ is a probability of event A given that event B has occurred. The equation can be written as seen here.

$$P(A|B) = \frac{P(A \cap B)}{P(B)} \tag{2.5}$$

For the reverse condition, $P(B|A)$, the equation can be written as seen here.

$$P(B|A) = \frac{P(A \cap B)}{P(A)} \tag{2.6}$$

Example 2.5 In a group of 100 defective mobile phone cover glass pieces, 40 pieces have scratch issues, 30 pieces have angle cutting issues, and 20 pieces have both scratch and angle cutting issues. If a defective phone cover glass piece, chosen at random, has a scratch issue, what is the probability that the item also has an angle cutting issue?

Answer:
Event A = scratch issue
Event B = angle cutting issue

$$P(B|A) = \frac{P(A \cap B)}{P(A)} = \frac{\frac{20}{100}}{\frac{40}{100}} = 0.5$$

2.5 Joint Probability with Dependence

If multiple events are dependent on each other, the joint probability can be calculated by considering the conditional probability:

$$P(AB) = P(A)P(B|A)$$

or

$$P(AB) = P(B)P(A \mid B) \tag{2.7}$$

Example 2.6 In a manufacturing plant, there are two main machines. For machine A, the probability of producing a defective part is 0.03. For machine B, the probability of seeing a defective part is 0.05. The two machines are dependent on each other. For example, if machine A produces a defective part, the probability of seeing a defective part from machine B is 0.06. What is the probability that both machines will produce defective parts simultaneously?

Answer:
Event A = a defective part from machine A
Event B = a defective part from machine B

$$P(AB) = P(A)P(B \mid A) = 0.03 \times 0.06 = 0.0018$$

2.6 Mutually Exclusive Events

If multiple events cannot occur simultaneously, they are called mutually exclusive events. For example, if we toss a coin, the upside coin can be either a head or a tail but not both. As seen in Figure 2.3, there is no intersection between events A and B. Thus, the joint probability, $P(AB)$, is 0. The union probability can be written as

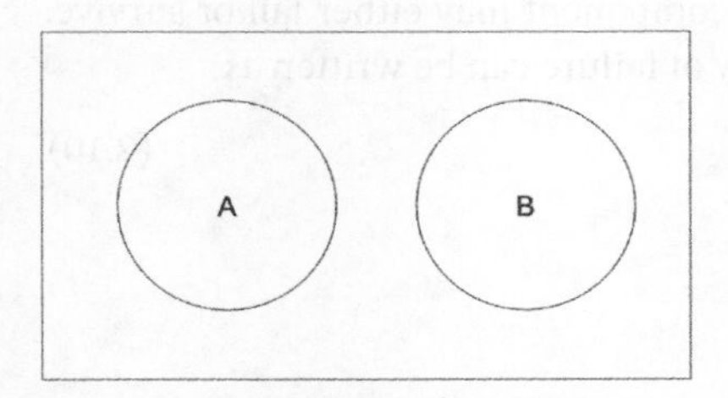

Figure 2.3 Mutually exclusive events.

$$P(A \cup B) = P(A) + P(B) \tag{2.8}$$

Example 2.7 In a group of 100 defective mobile phone cover glass pieces, 40 pieces have scratch issues, and 30 pieces have angle cutting issues. These two issues cannot happen together in one piece. What is the probability that a mobile phone cover glass piece has a scratch issue or an angle cutting issue?

Answer:
Event A = scratch issue
Event B = angle cutting issue

$$P(A \cup B) = P(A) + P(B) = \frac{40}{100} + \frac{30}{100} = 0.7$$

2.7 Complement Rule

If multiple events are exhaustive, at least one event must occur. This meets the complement rule as seen in Figure 2.4. With the complement rule, the union probability can be written as

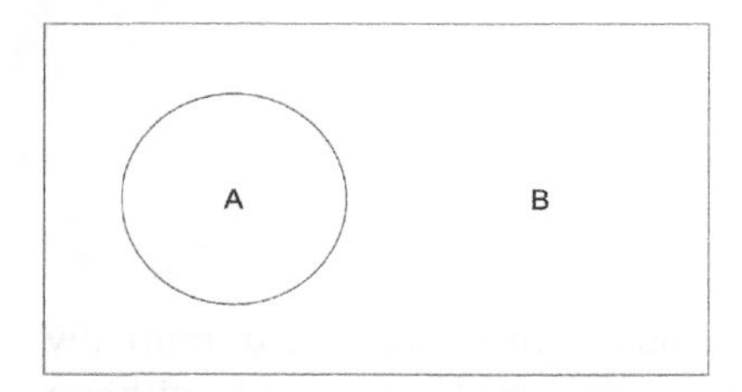

Figure 2.4 Complement rule.

$$P(A \cup B) = P(A) + P(B) = 1 \tag{2.9}$$

This rule can be applied to reliability. A component may either fail or survive. In this exhaustive condition, the probability of failure can be written as

$$P(\text{Failure}) = 1 - P(\text{Survival}) \tag{2.10}$$

2.8 Total Probability

The total probability rule is to find a probability of an event by summing up the probability of distinct parts. In Figure 2.5, if we are interested in calculating the probability of event C, the total probability rule can be considered. The formula could be written as

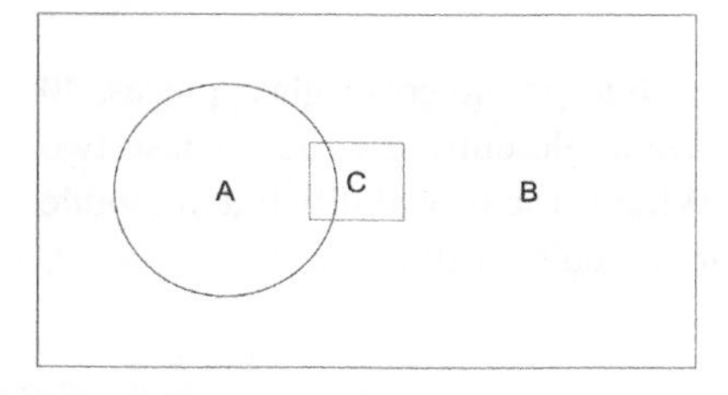

Figure 2.5 Total probability.

$$P(C) = P(CA) + P(CB) = P(A)P(C \mid A) + P(B)P(C \mid B) \quad (2.11)$$

In order to use the total probability rule, there are required assumptions. Based on Figure 2.5, the situation needs to meet the following:

$$(A \cup B) = 1$$

$$P(A \cap B) = 0 \quad (2.12)$$

Example 2.8 Three different products were produced in a manufacturing plant. Product A accounted for 60% of total products, product B was 30% of the total, and product C was 10%. Within the product A group, 10% failed, and products B and C groups had failure portions of 15% and 5%, respectively. If a product in the plant is randomly selected, what is the probability of it being a failed product?

Answer:
Event A = product A
Event B = product B
Event C = product C
Event F = failed product

$$P(A) = 0.6,\ P(B) = 0.3,\ P(C) = 0.1$$

$$P(F \mid A) = 0.1,\ P(F \mid B) = 0.15,\ P(F \mid C) = 0.05$$

$$\begin{aligned} P(F) &= P(AF) + P(BF) + P(CF) \\ &= P(A)P(F \mid A) + P(B)P(F \mid B) + P(C)P(F \mid C) \\ &= 0.6 \times 0.1 + 0.3 \times 0.15 + 0.1 \times 0.05 = 0.11 \end{aligned}$$

2.9 Bayes' Rule

Bayes' rule (also known as Bayes' theorem) is used to calculate the conditional probability based on prior knowledge of the conditions that could be related to the event. For instance, if we already know $P(B \mid A)$, can we determine $P(A \mid B)$?

The rule can be written as

$$P(A \mid B) = \frac{P(AB)}{P(B)} = \frac{P(A)P(B \mid A)}{P(B)} \quad (2.13)$$

If we know the probability of event B, the total probability rule can be applied.

$$P(A \mid B) = \frac{P(A)P(B \mid A)}{P(A)P(B \mid A) + P(A')P(B \mid A')} \quad (2.14)$$

Example 2.9 Three different products were produced in the manufacturing plant. Product A accounted for 60% of the total products, product B was 30% of the total, and product C was 10%. Within the product A group, 10% failed, and products B and C groups had failure portions of 15% and 5%, respectively. If a component fails, what is the probability that it is related to product A?

Answer:
Event A = product A
Event B = product B
Event C = product C
Event F = failed product

$$P(\mathrm{A}) = 0.6,\ P(\mathrm{B}) = 0.3,\ P(\mathrm{C}) = 0.1$$

$$P(\mathrm{F}\mid\mathrm{A}) = 0.1,\ P(\mathrm{F}\mid\mathrm{B}) = 0.15,\ P(\mathrm{F}\mid\mathrm{C}) = 0.05$$

$$\begin{aligned} P(\mathrm{F}) &= P(\mathrm{AF}) + P(\mathrm{BF}) + P(\mathrm{CF}) \\ &= P(\mathrm{A})P(\mathrm{F}\mid\mathrm{A}) + P(\mathrm{B})P(\mathrm{F}\mid\mathrm{B}) + P(\mathrm{C})P(\mathrm{F}\mid\mathrm{C}) \\ &= 0.6 \times 0.1 + 0.3 \times 0.15 + 0.1 \times 0.05 = 0.11 \end{aligned}$$

$$P(\mathrm{A}\mid\mathrm{F}) = \frac{P(\mathrm{F}\mid\mathrm{A})P(\mathrm{A})}{P(\mathrm{F})} = \frac{0.1 \times 0.6}{0.11} = 0.545$$

2.10 Summary

- Probability is a ratio of specific outcomes to total possible outcomes.
- Joint probability is the probability of multiple events occurring simultaneously.
- Union probability is the probability that either of multiple events may occur.
- Conditional probability is the probability of an event given that another event has occurred.
- If multiple events are dependent on each other, the joint probability can be calculated by considering the conditional probability.
- If multiple events cannot occur simultaneously, they are called mutually exclusive events.
- If multiple events are exhaustive, at least one event must occur. This meets the complement rule.
- The total probability rule says that we find a probability of an event by summing up the probability of the distinct parts.
- Bayes' rule is used to calculate the conditional probability based on prior knowledge of the conditions that could be related to the event.

Exercises

1 An inspection is regularly performed in a tire factory before shipping a product to customers. The total number of tires in a stack is 52. It is empirically known that the number of defective tires is 3. What is the probability of finding a defective tire from a tire stack?

2 Describe the required assumptions under the total probability rule.

3 Explain the formula of joint probability when multiple events are independent of each other.

4 Suppose that an assembly company has just completed the production of 100 parts. However, 10 of the parts were found defective. The first part out of the 100 was picked randomly, and set aside. The second part was selected randomly. What is the probability that both picked parts are defective?

5 Company A sells three brands of washing machines: brand A, brand B, and brand C. Of the machines they sell, 25% are brand A, 50% are brand B, and 25% are brand C. Based on past experience, the company's reliability engineers expect that the brand A machines will need service repairs with probability 0.1, brand B machines with probability 0.15, and brand C machines with probability 0.2. Find the overall probability that a customer will need service repairs on the washing machine they purchased.

3

Lifetime Distributions

Chapter Overview and Learning Objectives

- To understand the basic concepts of the probability distributions.
- To learn the discrete and continuous probability distributions.
- To apply the probability distributions to the reliability concepts.
- To understand the fundamental concepts of the failure rate.
- To learn parameters and failure characteristics of various distributions, including exponential, Weibull, normal, and lognormal distributions.
- To apply Excel, Minitab, and Python functions to efficiently calculate the values related to the probability distributions.

3.1 Probability Distributions

The probability distribution gives the probabilities of occurrences of different possible outcomes. Figure 3.1 shows the probability distribution of the number of cars sold per day. The highest probability is to sell three cars per day, and the lowest probability is to sell five cars per day.

3.1.1 Random Variables

A random variable is an outcome from a random experiment. These values could vary depending on the results of a random experiment. The random variable can be categorized into discrete and continuous variables.

- Discrete random variable
  A discrete random variable accounts for countable values or distinct values. A finite number of distinct values is expected. Examples include:

Reliability Analysis Using MINITAB and Python, First Edition. Jaejin Hwang.
© 2023 John Wiley & Sons, Inc. Published 2023 by John Wiley & Sons, Inc.
Companion Website: www.wiley.com\go\Hwang\ReliabilityAnalysisUsingMinitabandPython

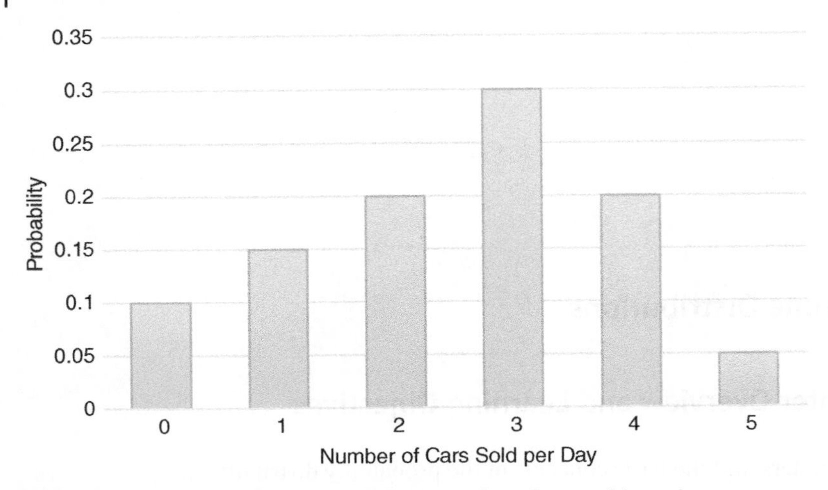

Figure 3.1 Probability distribution of the number of cars sold per day.

  - Number of defective items in a lot
  - Number of good products sold to customers
  - Number of errors in a software program
- Continuous random variable
  A continuous random variable could take infinitely many values. It is typically obtained from a measurement. Examples include:
  - The time to failure of a product
  - The length of a part
  - The weight of a component

3.2 Discrete Probability Distribution

A discrete probability distribution accounts for discrete random variables that have probability values at discrete points.

- Probability mass function (PMF)
  A probability mass function shows the magnitude of the probability of each discrete point.
- Cumulative distribution function (CDF)
  A cumulative distribution function is the probability that a random variable (X) will take a value less than or equal to a certain value (x).

Example 3.1 The number of defective items at a plant is a random variable (X). The probability distribution for X is summarized in Table 3.1 as a probability mass function (PMF). Create a table showing a cumulative distribution function (CDF).

Table 3.1 A probability mass function of the number of defective items (X).

X	$P(X)$
0	0.3
1	0.2
2	0.2
3	0.2
4	0.1

Answer:

X	$P(X \leq x)$
0	$P(X \leq 0) = 0.3$
1	$P(X \leq 1) = 0.5$
2	$P(X \leq 2) = 0.7$
3	$P(X \leq 3) = 0.9$
4	$P(X \leq 4) = 1.0$

Example 3.2 The number of defective items at a plant is a random variable (X). The probability distribution for X is summarized in Table 3.2.

Table 3.2 The probability distribution for X.

X	$P(X)$
0	0.3
1	0.2
2	0.2
3	0.2
4	0.1

Find the probability of:

1) Exactly 3 defective items
2) At least 1 defective item
3) At most 2 defective items

Answer:

1) Exactly 3 defective items

$$P(X=3)=0.2$$

2) At least 1 defective item

$$P(X\geq 1)=0.2+0.2+0.2+0.1=0.7$$

Or

$$1-P(X<1)=1-0.3=0.7$$

3) At most 2 defective items

$$P(X\leq 2)=0.4+0.2+0.2=0.8$$

3.3 Continuous Probability Distribution

A continuous probability distribution shows the probabilities that are areas under the curve (integrals). Because infinite values can be assumed for a random variable, the probability of a random variable on any particular value is zero. Instead, the probabilities are provided for a range of values.

- Probability density function (PDF)
  The probability density function is a derivative of the cumulative distribution function. It shows the random variable's (X's) probability within a range of values.

$$P(a\leq X\leq b)=\int_a^b f(x)dx \tag{3.1}$$

- Cumulative distribution function (CDF)
  The cumulative distribution function provides the probability of the random variable at a value less than or equal to a certain value.

$$F(x)=P(X\leq a)=\int_0^a f(x)dx \tag{3.2}$$

The total area under the probability density function is 1.

$$\int_{-\infty}^{+\infty} f(x)dx = 1 \tag{3.3}$$

Since $P(X = x)$ is zero, various ranges can be expressed as:

$$P(a < X < b) = P(a \leq X < b) = P(a < X \leq b) = P(a \leq X \leq b) \tag{3.4}$$

3.3.1 Reliability Concepts

A continuous probability distribution can be applied to reliability. A random variable commonly used is **time to failure**. Figure 3.2 shows the probability density function of the time to failure.

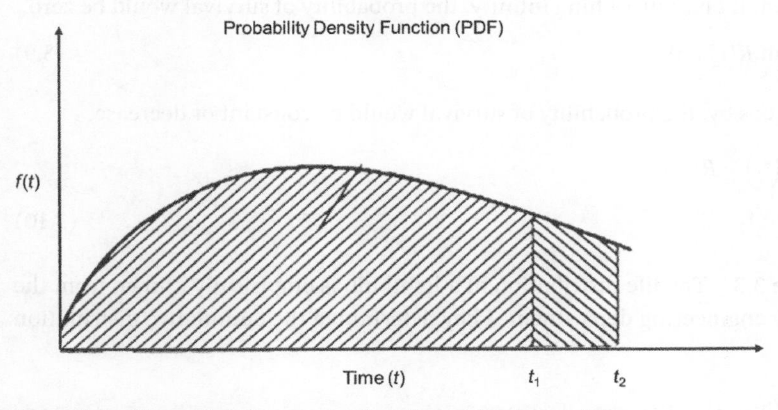

Figure 3.2 The probability density function of the time to failure.

The cumulative distribution function can be written as:

$$F(t) = P(X \leq t) = \int_0^t f(x)dx \tag{3.5}$$

This indicates the probability that a randomly selected component would fail by t (hours, week, years, or etc.). It can also be interpreted that a portion of all components in the population would fail by t.

Based on Figure 3.2, the probability of the failure in a certain interval of time could be written as:

$$F(t_2) - F(t_1) = P(X \leq t_2) - P(X \leq t_1) = \int_{t_1}^{t_2} f(x)dx \tag{3.6}$$

This indicates the probability that a component would survive to t_1 but fail by t_2. It could also be interpreted that a portion of the population would fail in that interval.

The reliability or survival function can be considered using the continuous probability distribution.

$$R(t)=1-F(t)=1-P(X\leq t)=P(X>t)=\int_{t}^{\infty}f(x)dx \tag{3.7}$$

This indicates the probability that a randomly selected part would still operate after t. It can also be interpreted that a portion of components in the population would survive to at least t.

Several characteristics would be expected for the reliability function. If t is zero, the probability of survival should be expected to be 1.

$$R(0)=1 \tag{3.8}$$

In contrast, if t is approaching infinity, the probability of survival would be zero.

$$\lim_{t\to\infty}R(t)=0 \tag{3.9}$$

As time goes by, the probability of survival would be constant or decrease.

$$R(t_1)\geq R(t_2)$$

$$t_1<t_2 \tag{3.10}$$

Example 3.3 The life distribution of a lightbulb is empirically known from the reliability engineering department. They determined the cumulative distribution function:

$$P(X\leq t)=F(t)=1-(1+0.005\times t)^{-1}$$

1) What is the probability that a bulb will fail by 500 hours?
2) What proportion of bulbs will survive more than 1000 hours?
3) What proportion of bulbs will last more than 1000 hours but fail by 1500 hours?

Answer:

1) What is the probability that a bulb will fail by 500 hours?

$$P(X\leq 500)=F(500)=1-(1+0.005\times 500)^{-1}=0.714$$

2) What proportion of bulbs will survive more than 1000 hours?

$$P(X>1000)=1-F(1000)=(1+0.005\times 1000)^{-1}=0.167$$

3) What proportion of bulbs will last more than 1000 hours but fail by 1500 hours?

$$P(1000 < X \leq 1500) = F(1500) - F(1000)$$
$$= \left[1-(1+0.005\times 1500)^{-1}\right] - \left[1-(1+0.005\times 1000)^{-1}\right] = 0.049$$

3.3.2 Failure Rate

The failure rate (or hazard rate) is the number of failures per unit of time. The failure rate concept was used to describe the reliability bathtub curve. The conditional probability rules can be applied to describe the failure rate.

$$P(\text{B}\,|\,\text{A}) = \frac{P(\text{AB})}{P(\text{A})}$$

$$P(\text{fail in next } \Delta t \,|\, \text{survive to time } t) = \frac{F(t+\Delta t)-F(t)}{R(t)} \tag{3.11}$$

Equation 3.11 indicates a probability that a component would fail in the next unit of time given it has survived for t. We divide this conditional probability by Δt to convert it to a failure rate.

$$\frac{F(t+\Delta t)-F(t)}{R(t)\Delta t} \tag{3.12}$$

If we assume that Δt would approach zero, the result would be this equation.

$$F'(t) = \lim_{\Delta t \to 0} \frac{F(t+\Delta t)-F(t)}{\Delta t}$$

$$\frac{F(t+\Delta t)-F(t)}{R(t)\Delta t} = \frac{F'(t)}{R(t)} = \frac{f(t)}{R(t)}$$

$$h(t) = \frac{f(t)}{R(t)} = \frac{f(t)}{1-F(t)} \tag{3.13}$$

Failure rates per hour unit are often very small numbers to interpret. There are two common units used for the failure rate.

- Percent per thousand hours (%/K):
  It indicates a rate of failure for every 100 items operating for 1000 hours.
- Fails in Time (FIT):
  It shows a rate of failure for 10^9 operating hours.

Example 3.4 There is a constant failure rate of 0.1 after year 2. Based on this information, fill out Table 3.3.

Table 3.3 The failure characteristics with a constant failure rate of 0.1 after year 2.

	Year 1	Year 2	Year 3	Year 4	Year 5
Number of survivals	100				
Number of failures	0				
Failure rate	0	0.1	0.1	0.1	0.1
Number of cumulative failures	0				

Answer:

	Year 1	Year 2	Year 3	Year 4	Year 5
Number of survivals	100	90	81	73	66
Number of failures	0	10	9	8	7
Failure rate	0	0.1	0.1	0.1	0.1
Number of cumulative failures	0	10	19	27	34

Example 3.5 A life test was conducted on 1000 components. As a result, 100 items failed at between 100 and 110 hours. The number of components that survived by 100 hours was 600. Calculate the failure rate in this group.

Answer:

$$h(t) = \frac{F(t+\Delta t) - F(t)}{R(t)\Delta t} = \frac{\frac{100}{1000}}{\frac{600}{1000} \times 10} = 0.0167 \text{ failure per hour}$$

Example 3.6 The life distribution of a lightbulb is empirically known from the reliability engineering department. They determined the cumulative distribution function:

$$P(X \leq t) = F(t) = 1 - (1 + 0.005 \times t)^{-1}$$

Derive a failure rate equation, and calculate the failure rate at 500 hours in both %/K and FIT units.

Answer:

$$f(t) = F'(t) = 0.005(1 + 0.005 \times t)^{-2}$$

$$h(t) = \frac{f(t)}{1 - F(t)} = \frac{0.005(1 + 0.005 \times t)^{-2}}{(1 + 0.005 \times t)^{-1}} = 0.005(1 + 0.005 \times t)^{-1}$$

$$h(500) = 0.005(1 + 0.005 \times 500)^{-1} = 0.001429$$

$$0.001429 \times 10^5 = 142.9\% / K$$

$$0.001429 \times 10^9 = 1429000 \ FITs$$

3.4 Exponential Distribution

The exponential distribution is one of the widely used continuous distributions. The unique feature of the exponential distribution is that it has a constant failure rate function, which means that the failure rate is not affected by time. An exponential distribution is a simple distribution due to the ease of mathematical calculations. In the reliability field, an exponential distribution is commonly applied to high-quality integrated circuits such as diodes, transistors, resistors, and capacitors.

A single parameter model will be discussed in this book. Here is an equation of the probability density function:

$$f(t) = \lambda e^{-\lambda t}$$

$$t \geq 0, \ \lambda \geq 0 \tag{3.14}$$

A lambda (λ) represents a constant failure rate, and functions as a scale parameter. A variable t indicates a time to failure. Figure 3.3 illustrates the probability density function of the exponential distribution.

The cumulative distribution function of the exponential distribution can be written as

$$F(t) = 1 - e^{-\lambda t}$$

$$t \geq 0, \ \lambda \geq 0 \tag{3.15}$$

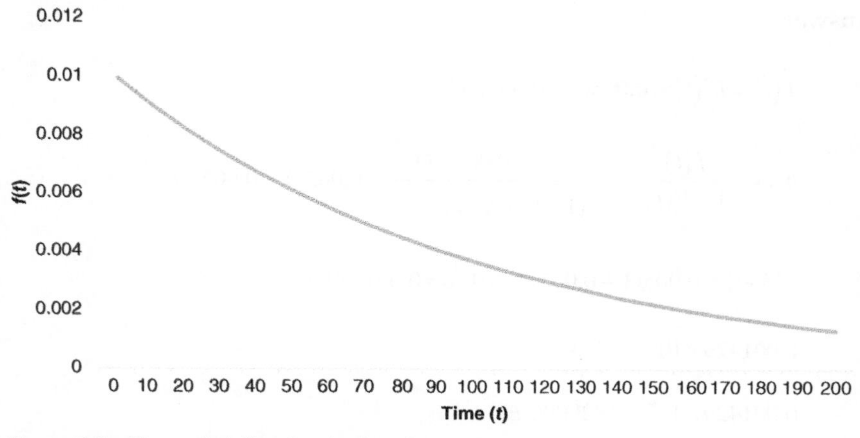

Figure 3.3 The probability density function of the exponential distribution.

Figure 3.4 illustrates the cumulative distribution function of the exponential distribution.

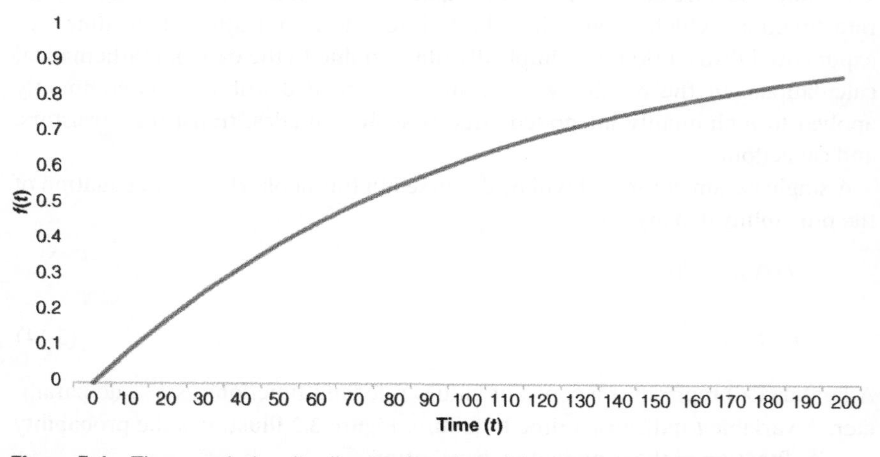

Figure 3.4 The cumulative distribution function of the exponential distribution.

Example 3.7 A group of diodes have a constant failure rate of 0.002 failures per hour. What is the probability that the diodes will fail by 1000 hours?

Answer:

$$F(1000) = 1 - e^{-0.002 \times 1000} = 0.865$$

The mean time to failure (MTTF) measures the average amount of time to the failures of components. This is closely related to the lambda (λ) parameter.

$$\mathrm{MTTF} = \frac{1}{\lambda} \tag{3.16}$$

Example 3.8 Calculate the probability of failure when the time t reached the mean time to failure (MTTF).

Answer:

$$F(t) = 1 - e^{-\lambda t}$$

$$F(\mathrm{MTTF}) = 1 - e^{-\lambda(\mathrm{MTTF})} = 1 - e^{-\lambda\left(\frac{1}{\lambda}\right)} = 1 - e^{-1} = 0.632$$

The exponential reliability function can be written as:

$$R(t) = 1 - F(t) = 1 - \left(1 - e^{-\lambda t}\right) = e^{-\lambda t} \tag{3.17}$$

Figure 3.5 illustrates the exponential reliability function.

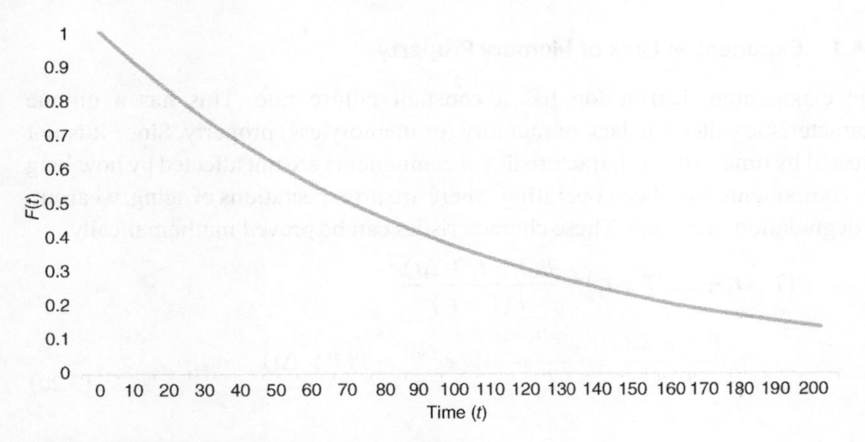

Figure 3.5 Exponential reliability function.

The exponential failure rate function can be written as

$$h(t) = \frac{f(t)}{R(t)} = \lambda \tag{3.18}$$

This indicates that the failure rate is a constant value and is not affected by time.

The time to failure can be calculated by inverting the formula of the cumulative distribution function:

$$t = \frac{-\ln\left(1 - F(t)\right)}{\lambda} \tag{3.19}$$

Example 3.9 Transistors are known to fail related to the exponential distribution. Based on historical data, reliability engineers determined the failure rate as 0.20%/K.

1) Calculate the MTTF.
2) Calculate the probability that transistors will last more than 15,000 hours.
3) Calculate the point in time that 10% of transistors will fail.

Answer:

1) Calculate the MTTF.

$$\text{MTTF} = \frac{1}{\lambda} = \frac{1}{0.20 \times 10^{-5}} = 500{,}000 \text{ hours}$$

2) Calculate the probability that transistors will last more than 15,000 hours.

$$R(15000) = e^{-\lambda t} = e^{-0.2 \times 10^{-5} \times 15000} = 0.970$$

3) Calculate the point in time that 10% of transistors will fail.

$$t = \frac{-\ln(1 - F(t))}{\lambda} = \frac{-\ln(1 - 0.1)}{0.2 \times 10^{-5}} = 52{,}680 \text{ hours}$$

3.4.1 Exponential Lack of Memory Property

The exponential distribution has a constant failure rate. This has a unique characteristic called the lack of memory (or memoryless) property. Since it is not affected by time, failure characteristics of components are not affected by how long the components have been operating. There are no expectations of aging, wearout, or degradation over time. These characteristics can be proved mathematically.

$$P(T > t_1 + \Delta t \mid T > t_1) = \frac{P(T > t_1 + \Delta t)}{P(T > t_1)}$$

$$= \frac{R(t_1 + \Delta t)}{R(t_1)} = \frac{e^{-\lambda(t_1 + \Delta t)}}{e^{-\lambda t_1}} = e^{-\lambda \Delta t} = P(T > \Delta t) \tag{3.20}$$

Example 3.10 The failure characteristics of capacitors have an exponential distribution with a mean time to failure of 2 years.

1) Calculate the probability that capacitors will fail by the first 0.5 year.
2) Suppose that the capacitors survived by 1 year. Calculate the probability that capacitors will fail in the next 0.5 year.

Answer:

1) Calculate the probability that capacitors will fail by the first 0.5 year.

$$P(X \leq 0.5) = F(0.5) = 1 - e^{-\frac{1}{2} \times 0.5} = 0.221$$

2) Suppose that capacitors survived by 1 year. Given this condition, calculate the probability that the capacitors will fail in the next 0.5 year.

$$P\left(X \le 1.5 \mid X > 1\right) = \frac{P(1 < X \le 1.5)}{P(X > 1)} = \frac{F\left(1.5\right) - F\left(1\right)}{R\left(1\right)} = \frac{\left[1 - e^{-\frac{1}{2}\times 1.5}\right] - \left[1 - e^{-\frac{1}{2}\times 1}\right]}{e^{-\frac{1}{2}\times 1}}$$

$$= 0.221$$

3.4.2 Excel Practice

The Excel functions can be used to efficiently calculate the exponential distribution.

- EXPONDIST(t, λ, TRUE)
  - TRUE = cumulative distribution function
  - FALSE = probability density function

3.4.3 Minitab Practice

The PDF with varying parameter values of the exponential distribution can be constructed using Minitab. Let's assume that we want to compare four different lambda values: $\lambda = 0.001, 0.002, 0.003, 0.004$.

Go to Graph > Probability Distribution Plot (Figure 3.6).

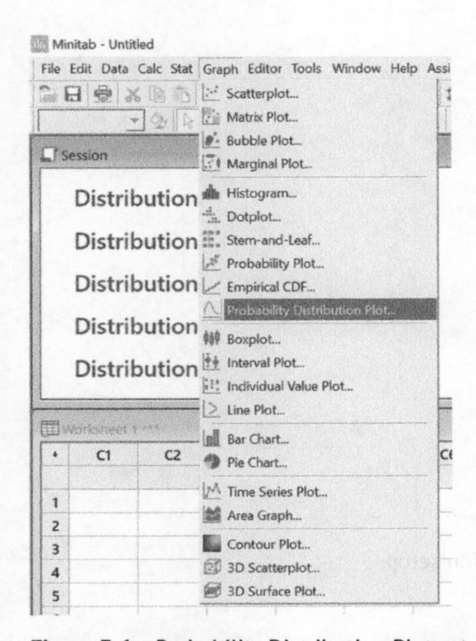

Figure 3.6 Probability Distribution Plot setup with Minitab.

Select [Vary Parameters] (Figure 3.7).

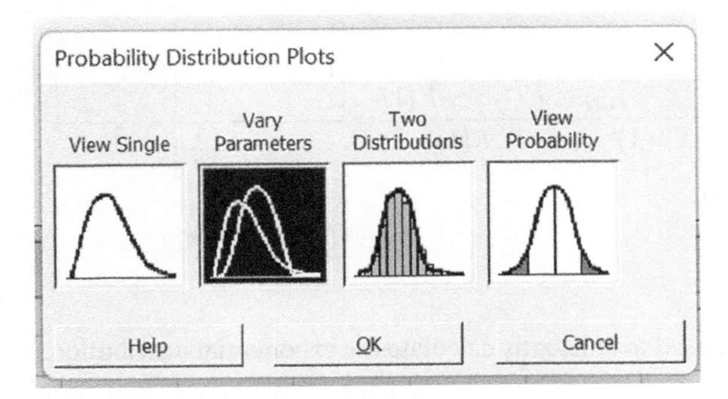

Figure 3.7 Vary Parameters function.

Choose the [Exponential] distribution. [Scales] indicates the MTTF. If we are interested in $\lambda = 0.001, 0.002, 0.003, 0.004$, MTTF = 1000, 2000, 3000, and 4000 (Figure 3.8). The threshold is to shift the distribution. This can be used for the two-parameter exponential distribution. In this example, we assume there is no shift in the distribution.

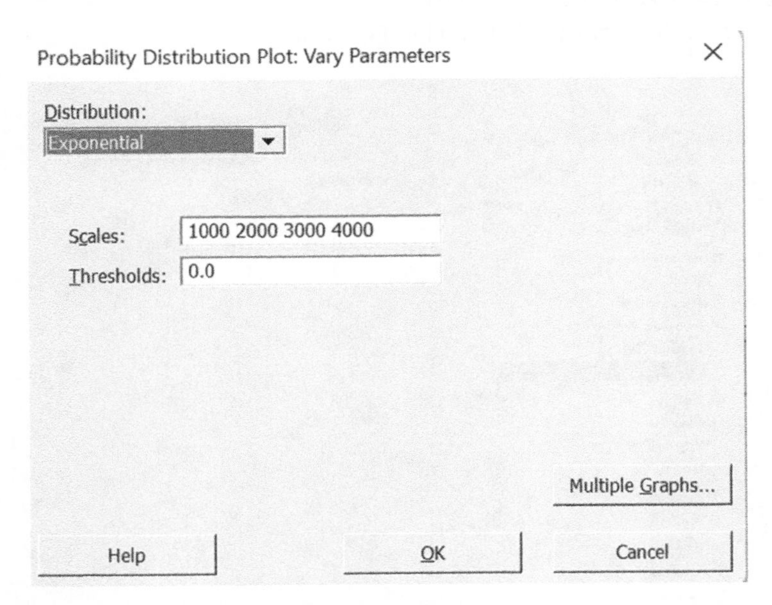

Figure 3.8 PDF of the exponential distribution setup.

The PDF with varying parameter values of the exponential distribution is constructed (Figure 3.9).

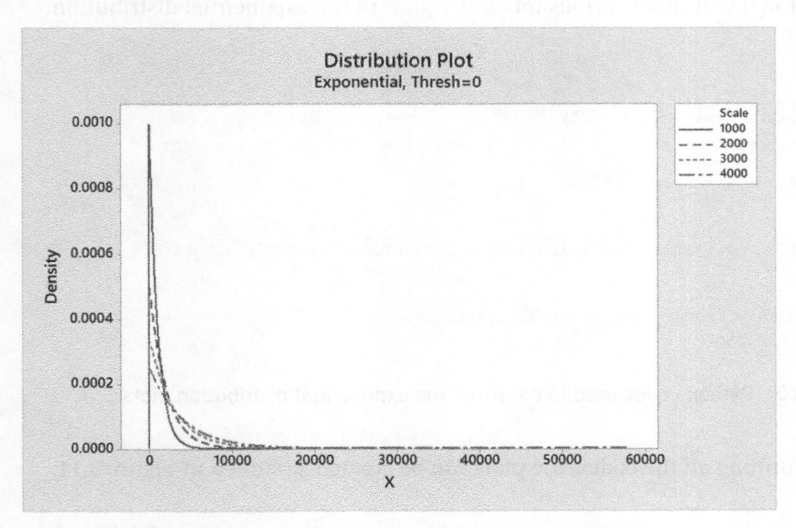

Figure 3.9 PDF of the exponential distribution with varying parameter values.

3.4.4 Python Practice

If we know the parameter information of the exponential distribution, PDF, CDF, $R(t)$, $h(t)$, and cumulative $h(t)$ plots can be constructed using Python. Let's assume that $\lambda = 0.001$.

Google Colab (https://colab.research.google.com) can be used to run the Python codes. The Python codes used to construct the exponential distribution plots are shown in Figure 3.10.

[pip install] would allow us to install the specific library resource. Here, we will install the reliability library for our analysis purpose.

We will also install [matplotlib] to create visualizations in Python.

There are eight different probability distributions available in [reliability. Distributions].

- Weibull distribution
- Exponential distribution
- Gamma distribution
- Normal distribution
- Lognormal distribution
- Loglogistic distribution
- Gumbel distribution
- Beta distribution

We will import [Exponential_Distribution] for our purpose.

We could assign the specific value of the parameter Lambda.

[dist.plot()] will show various reliability plots of the exponential distribution.

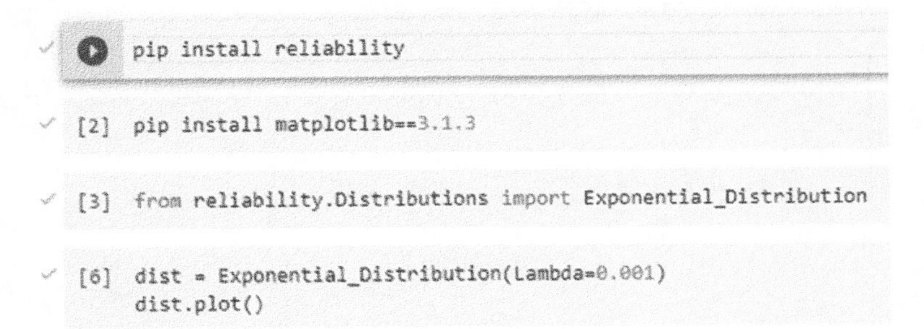

pip install reliability
[2] pip install matplotlib==3.1.3
[3] from reliability.Distributions import Exponential_Distribution
[6] dist = Exponential_Distribution(Lambda=0.001)
dist.plot()

Figure 3.10 Python codes used to construct the exponential distribution plots.

After running all the codes, the plots can be created as shown in Figure 3.11.

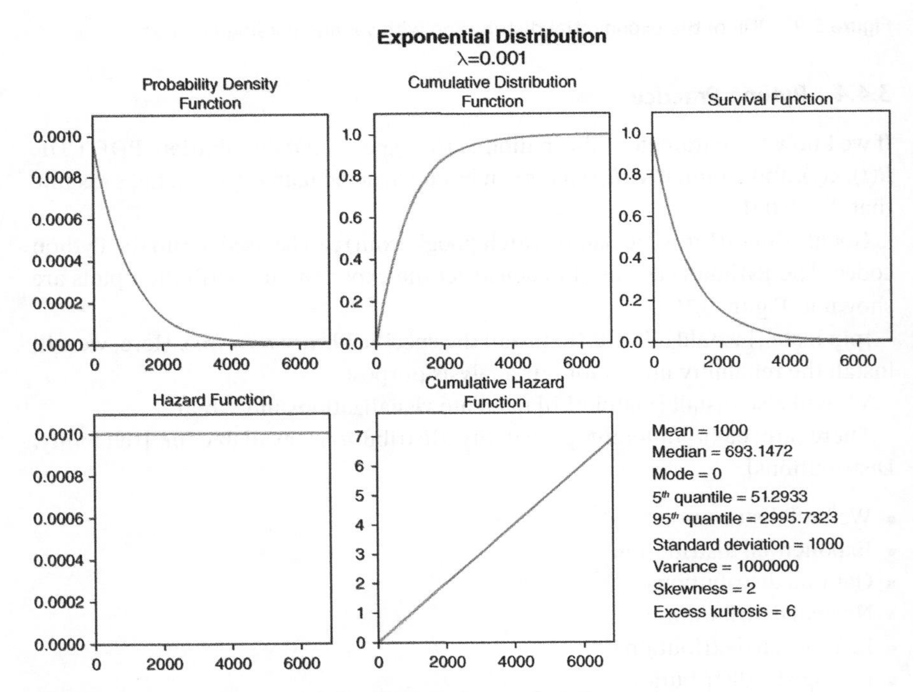

Figure 3.11 The exponential distribution plots with Python.

Tip! Crosshairs Function

The crosshairs function can be added to interactively explore the values in the plot. For the interactive GUI, IDLE platform can be used instead of Google Colab. The Python codes for an exponential CDF are shown in Figure 3.12.

[crosshairs] function is imported.

[xlabel] shows the time and [ylabel] shows a probability of cumulative failure.

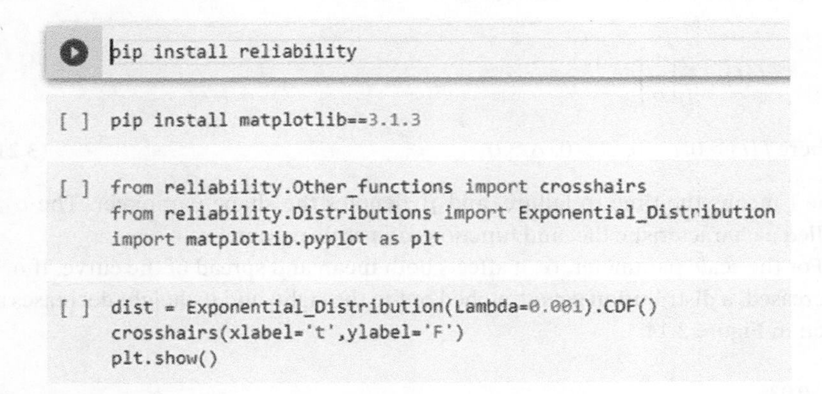

pip install reliability
[ ] pip install matplotlib==3.1.3
[ ] from reliability.Other_functions import crosshairs
from reliability.Distributions import Exponential_Distribution
import matplotlib.pyplot as plt
[ ] dist = Exponential_Distribution(Lambda=0.001).CDF()
crosshairs(xlabel='t',ylabel='F')
plt.show()

Figure 3.12 Python codes to implement the crosshairs function.

After running all the codes, the interactive exponential CDF can be constructed (Figure 3.13).

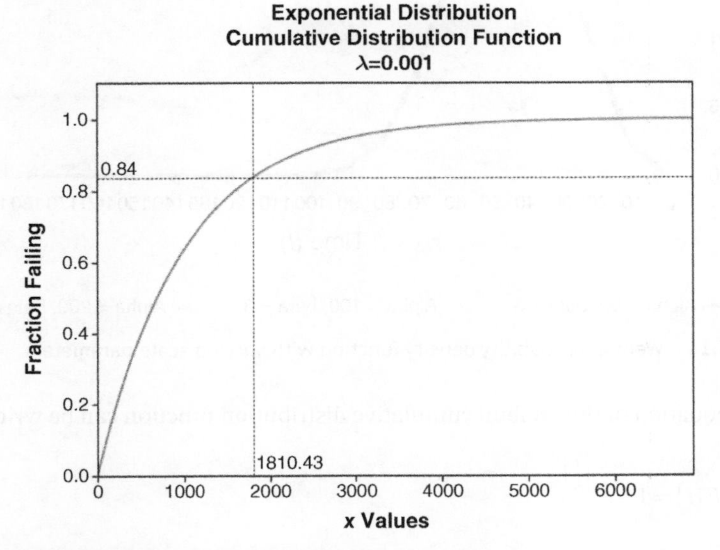

Figure 3.13 Exponential CDF with crosshairs function.

3.5 Weibull Distribution

The Weibull distribution was developed by Dr. Waloddi Weibull. This distribution is one of the most widely used lifetime distributions in reliability. The distribution is versatile, so it can be applied to increasing, constant, and decreasing failure rates over time.

The two-parameter model will be covered in this book. Here is an equation of the probability density function (PDF):

$$f(t)=\frac{\beta}{t}\left(\frac{t}{\alpha}\right)^{\beta}e^{-\left(\frac{t}{\alpha}\right)^{\beta}}$$

$$\text{where } f(t)\geq 0,\ t\geq 0,\ \beta>0,\ \alpha>0 \tag{3.21}$$

The t means the time to failure, and β denotes the shape parameter. The α is called a characteristic life, and functions as a scale parameter.

For the scale parameter, α, it affects both mean and spread of the curve. If α is increased, a distribution gets stretched out to the right, and its height decreases as seen in Figure 3.14.

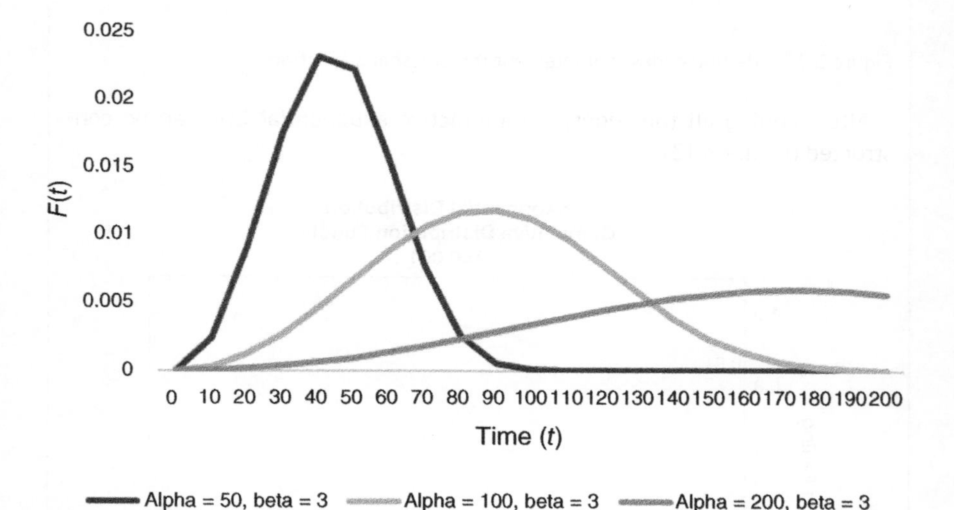

Figure 3.14 Weibull probability density function with varying scale parameters.

The equation of the Weibull cumulative distribution function can be written as

$$F(t)=1-e^{-\left(\frac{t}{\alpha}\right)^{\beta}} \tag{3.22}$$

Figure 3.15 illustrates the Weibull cumulative distribution function with varying scale parameters.

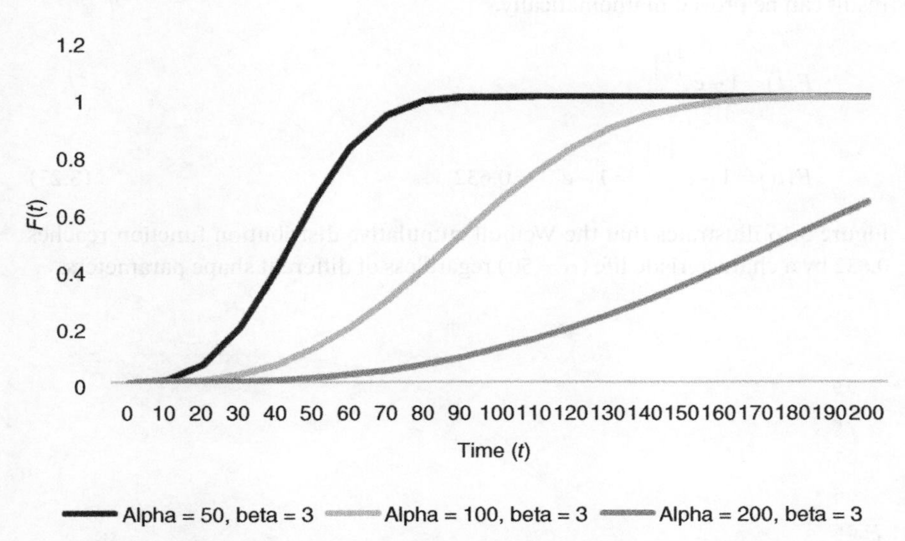

Figure 3.15 Weibull cumulative distribution function with varying scale parameters.

Example 3.11 A component's failure characteristics have a Weibull distribution with parameters $\beta = 10$ and $\alpha = 100$ days.

1) Calculate the probability that the component would fail by 105 days.
2) Calculate the probability of failure between 98 and 102 days.

Answer:

1) Calculate the probability that the component would fail by 105 days.

$$F(105) = 1 - e^{-\left(\frac{t}{\alpha}\right)^{\beta}} = 1 - e^{-\left(\frac{105}{100}\right)^{10}} = 0.804$$

2) Calculate the probability of failure between 98 and 102 days.

$$P(98 \leq X \leq 102) = F(102) - F(98) = \left[1 - e^{-\left(\frac{102}{100}\right)^{10}}\right] - \left[1 - e^{-\left(\frac{98}{100}\right)^{10}}\right] = 0.146$$

For the Weibull cumulative distribution function, 63.2% of the components would fail by a characteristic life, α, regardless of different parameters of β. This result can be proved mathematically.

$$F(t) = 1 - e^{-\left(\frac{t}{\alpha}\right)^{\beta}}$$

$$F(\alpha) = 1 - e^{-\left(\frac{\alpha}{\alpha}\right)^{\beta}} = 1 - e^{-1} = 0.632 \tag{3.23}$$

Figure 3.16 illustrates that the Weibull cumulative distribution function reaches 0.632 by a characteristic life ($\alpha = 50$) regardless of different shape parameters.

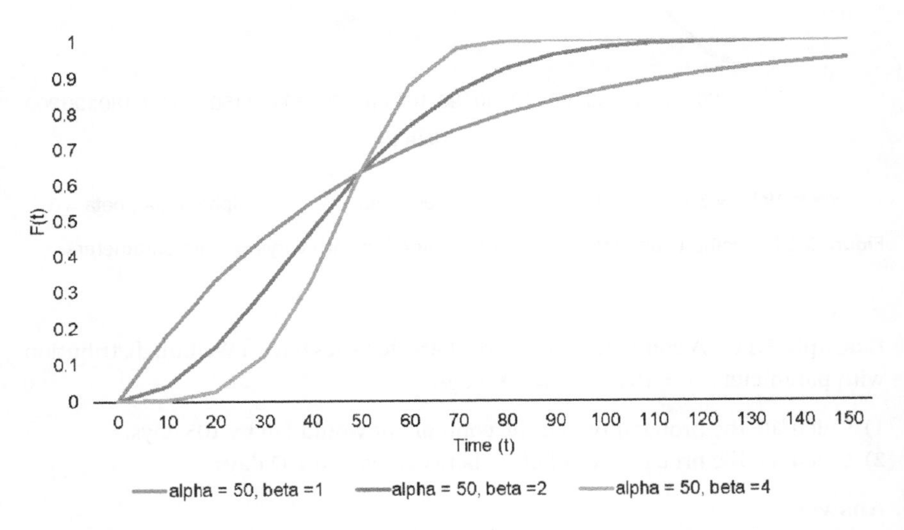

Figure 3.16 Weibull cumulative distribution function with varying shape parameters.

The equation of the Weibull reliability function can be written as

$$R(t) = 1 - F(t) = 1 - \left(1 - e^{-\left(\frac{t}{\alpha}\right)^{\beta}}\right) = e^{-\left(\frac{t}{\alpha}\right)^{\beta}} \tag{3.24}$$

Figure 3.17 illustrates the Weibull reliability function with varying shape parameters.

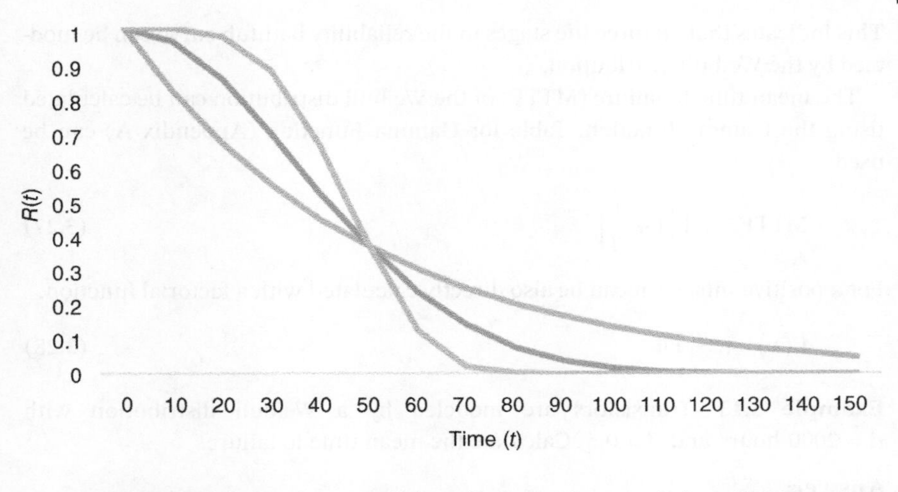

Figure 3.17 Weibull reliability function with varying shape parameters.

Example 3.12 Transistors are modeled by a Weibull distribution with $\alpha = 2000$ hours and $\beta = 0.5$. Calculate the reliability after 4000 hours.

Answer:

$$R(4000) = e^{-\left(\frac{4000}{2000}\right)^{0.5}} = 0.243$$

The equation of the Weibull failure rate function is

$$h(t) = \frac{f(t)}{R(t)} = \frac{\beta}{t}\left(\frac{t}{\alpha}\right)^{\beta} \tag{3.25}$$

There are several characteristics of the Weibull failure rate function depending on the shape parameter values.

- For $\beta < 1$, a failure rate decreases over time.
- For $\beta = 1$, there is a constant failure rate as an exponential distribution.

$$h(t) = \frac{\beta}{t}\left(\frac{t}{\alpha}\right)^{\beta} = \frac{1}{t}\left(\frac{t}{\alpha}\right)^{1} = \frac{1}{\alpha} \tag{3.26}$$

- For $\beta > 1$, a failure rate increases over time.
- For $\beta = 2$, there is a linear increase in the failure rate over time as a Rayleigh distribution.

This indicates that all three life stages in the reliability bathtub curve can be modeled by the Weibull distribution.

The mean time to failure (MTTF) of the Weibull distribution can be calculated using the Gamma function. Table for Gamma Function (Appendix A) can be used.

$$\text{MTTF} = \alpha\Gamma\left(1+\frac{1}{\beta}\right) \tag{3.27}$$

For a positive integer, it can be also directly calculated with a factorial function.

$$\Gamma(x) = (x-1)! \tag{3.28}$$

Example 3.13 Transistors are modeled by a Weibull distribution with $\alpha = 2000$ hours and $\beta = 0.5$. Calculate the mean time to failure.

Answer:

$$\text{MTTF} = 2000\Gamma\left(1+\frac{1}{0.5}\right) = 2000\Gamma(3) = 2000\times 2! = 4000 \text{ hours}$$

The median time of the Weibull distribution can be calculated as:

$$T_{\text{med}} = \alpha(\ln 2)^{\frac{1}{\beta}} \tag{3.29}$$

Example 3.14 Derive the equation of the median time of the Weibull distribution.

Answer:
It is expected that half of the population would fail by the median time.

$$R(T_{\text{med}}) = e^{-\left(\frac{T_{\text{med}}}{\alpha}\right)^{\beta}} = 0.5$$

$$\left(\frac{T_{\text{med}}}{\alpha}\right)^{\beta} = -\ln 0.5$$

$$T_{\text{med}} = \alpha(-\ln 0.5)^{\frac{1}{\beta}} = \alpha(\ln 2)^{\frac{1}{\beta}}$$

The time to failure can be calculated based on the cumulative distribution function.

$$t = \alpha\left[-\ln\left(1-F(t)\right)\right]^{\frac{1}{\beta}} \tag{3.30}$$

Example 3.15 Transistors are modeled by a Weibull distribution with $\alpha = 2000$ hours and $\beta = 0.5$. Calculate the time to achieve 20% failure.

Answer:

$$t = \alpha\left[-\ln\left(1-F(t)\right)\right]^{\frac{1}{\beta}} = 2000\left[-\ln\left(1-0.2\right)\right]^{\frac{1}{0.5}} = 99.586 \text{ hours}$$

The shape parameter can be calculated using the cumulative distribution function.

$$\beta = \frac{\ln\left[-\ln\left(1-F(t)\right)\right]}{\ln\frac{t}{\alpha}} \tag{3.31}$$

The scale parameter can be calculated using the cumulative distribution function.

$$\alpha = \frac{t}{[-\ln(1-F(t))]^{\frac{1}{\beta}}} \tag{3.32}$$

Example 3.16 Calculate the characteristic life necessary for 20% failures by 500 hours. A shape parameter is known to be 1.5.

Answer:

$$\alpha = \frac{500}{\left[-\ln\left(1-0.2\right)\right]^{\frac{1}{1.5}}} = 1359.087 \text{ hours}$$

The conditional probability rule can be applied to the Weibull distribution.

$$F\left(\text{future age}|\text{current age}\right) = 1 - e^{\left[\left(\frac{\text{current age}}{\alpha}\right)^{\beta} - \left(\frac{\text{future age}}{\alpha}\right)^{\beta}\right]} \tag{3.33}$$

Example 3.17 Transistors are modeled by a Weibull distribution with $\alpha = 2000$ hours and $\beta = 0.5$. Given that the transistors have survived for 3000 hours, calculate the probability that transistors will fail by the next 1000 hours.

Answer:

$$F\left(4000|3000\right) = 1 - e^{\left[\left(\frac{3000}{2000}\right)^{0.5} - \left(\frac{4000}{2000}\right)^{0.5}\right]} = 0.173$$

3.5.1 Excel Practice

The Excel functions can be used to efficiently calculate the Weibull distribution.

- WEIBULL(t, β, α, TRUE)
  - TRUE = cumulative distribution function
  - FALSE = probability density function

3.5.2 Minitab Practice

The PDF with varying parameter values of the Weibull distribution can be constructed using Minitab. Let's assume that we want to compare three different beta values, $\beta = 0.5, 1, 1.5$, with $\alpha = 2000$ hours.

Here is the setup of the PDF of the Weibull distribution (Figure 3.18). [Shapes] denotes β and [Scales] means α.

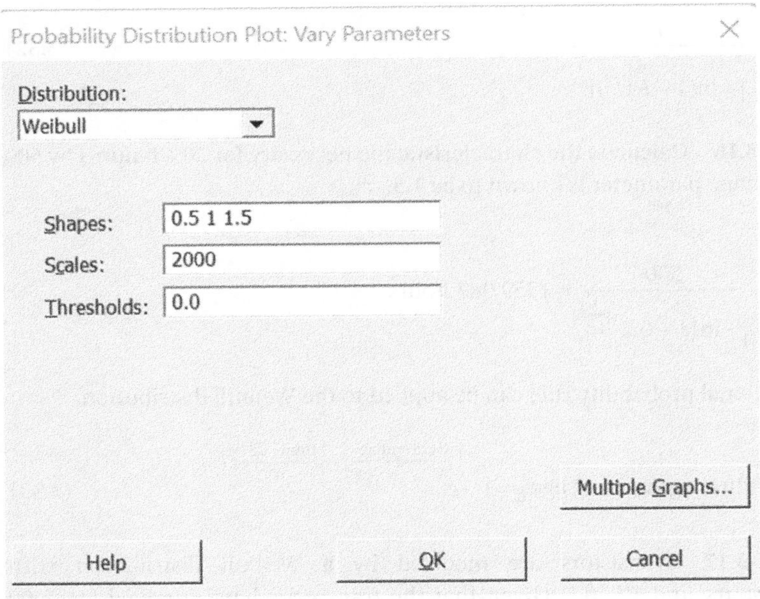

Probability Distribution Plot: Vary Parameters

Distribution:
Weibull

Shapes: 0.5 1 1.5
Scales: 2000
Thresholds: 0.0

Multiple Graphs...

Help OK Cancel

Figure 3.18 Weibull PDF with varying shape parameter values setup.

The PDF of the Weibull distribution with varying shapes is constructed as shown in Figure 3.19.

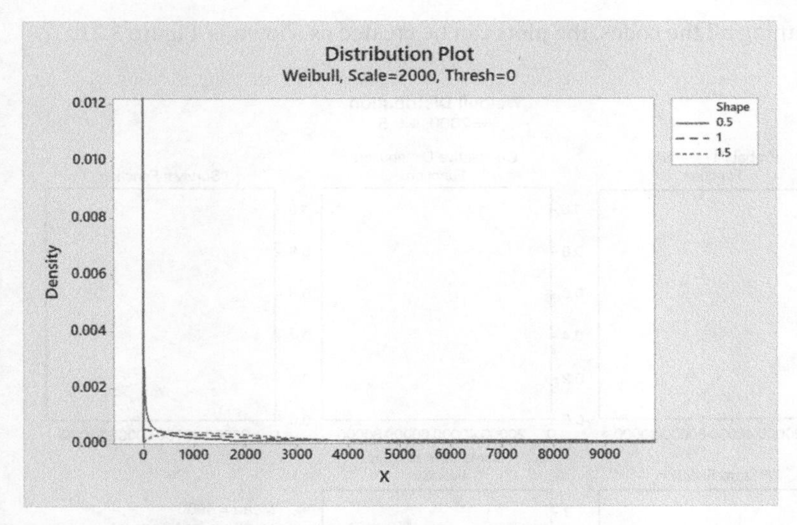

Figure 3.19 Weibull PDF with varying shape parameter values.

3.5.3 Python Practice

If we know the parameter information of the Weibull distribution, PDF, CDF, $R(t)$, $h(t)$, and cumulative $h(t)$ plots can be constructed using Python. Let's assume that $\alpha = 2000$ hours and $\beta = 0.5$.

The Python codes used to construct the Weibull distribution plots are shown in Figure 3.20.

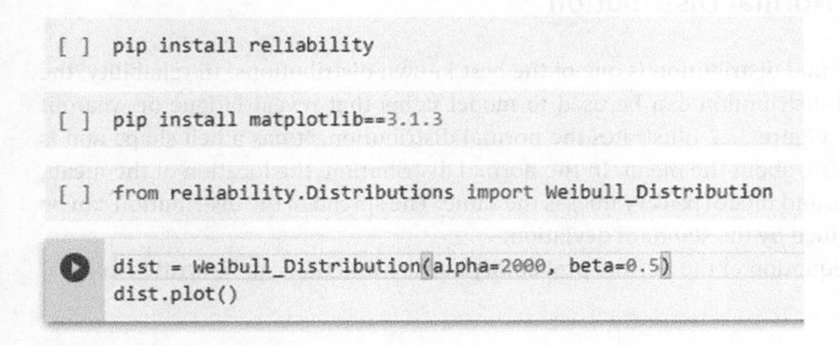

```
[ ] pip install reliability

[ ] pip install matplotlib==3.1.3

[ ] from reliability.Distributions import Weibull_Distribution

dist = Weibull_Distribution(alpha=2000, beta=0.5)
dist.plot()
```

Figure 3.20 Python codes used to construct the Weibull distribution plots.

After running all the codes, the plots can be created as shown in Figure 3.21.

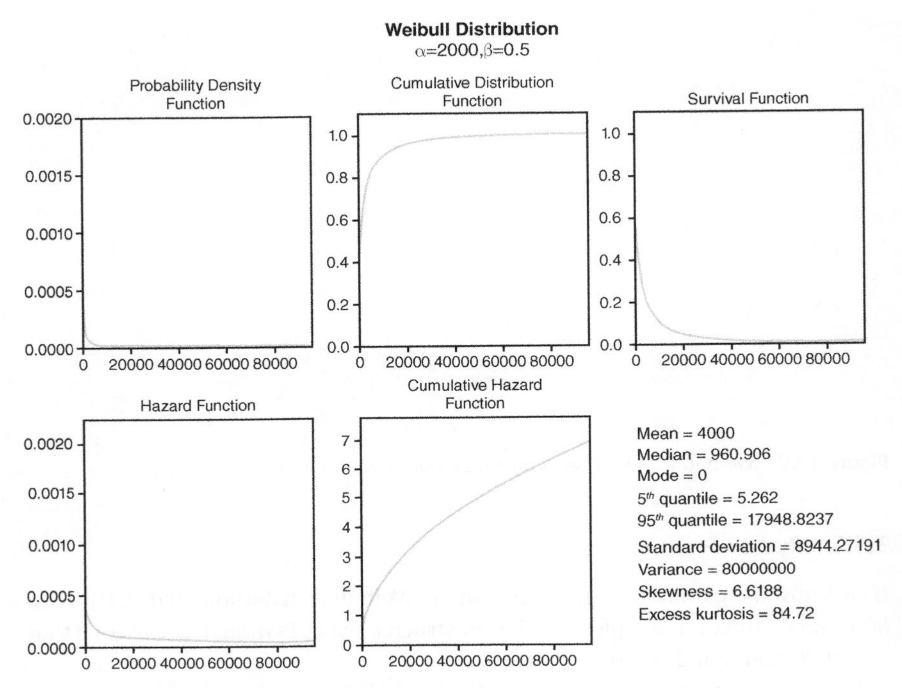

Figure 3.21 The Weibull distribution plots with Python.

3.6 Normal Distribution

The normal distribution is one of the best known distributions. In reliability, the normal distribution can be used to model items that reveal fatigue or wearout failure. Figure 3.22 illustrates the normal distribution. It has a bell shape and is symmetric about the mean. In the normal distribution, the location of the mean, median, and mode (peak) values is the same. The spread of the distribution can be determined by the standard deviation.

The equation of the normal probability density function can be written as

$$f(t)=\frac{1}{\sigma\sqrt{2\pi}}e^{-\frac{1}{2}\left(\frac{t-\mu}{\sigma}\right)^2} \tag{3.34}$$

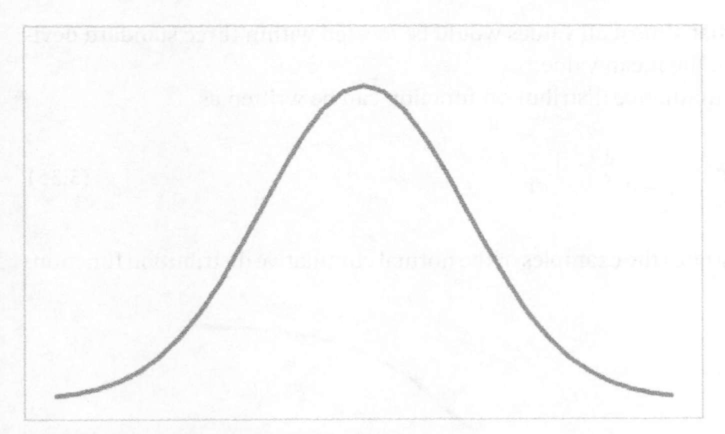

Figure 3.22 The normal distribution.

We consider a two-parameter model.

- μ is a mean time to failure, and it functions as a location parameter.
- σ is a standard deviation of the time to failure, and it functions as a scale parameter.

There is no shape parameter; the normal distribution has only a bell-shaped curve.

Regardless of different values of μ and σ, the 68-95-99.7 rule can be consistently applied to the normal distribution as seen in Figure 3.23.

- An area between $\mu \pm 1\sigma$ is 68.26% of the total area.
- An area between $\mu \pm 2\sigma$ is 95.44% of the total area.
- An area between $\mu \pm 3\sigma$ is 99.74% of the total area.

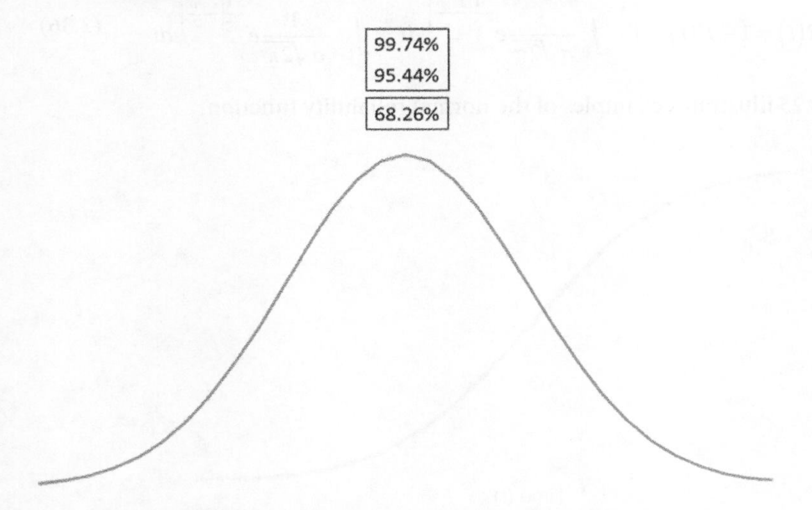

Figure 3.23 The 68-95-99.7 rule of the normal distribution.

This indicates that almost all values would be located within three standard deviations relative to the mean value.

The normal cumulative distribution function can be written as

$$F(t)=\int_0^t \frac{1}{\sigma\sqrt{2\pi}} e^{-\frac{1}{2}\left(\frac{t-\mu}{\sigma}\right)^2} dt \tag{3.35}$$

Figure 3.24 illustrates the examples of the normal cumulative distribution function.

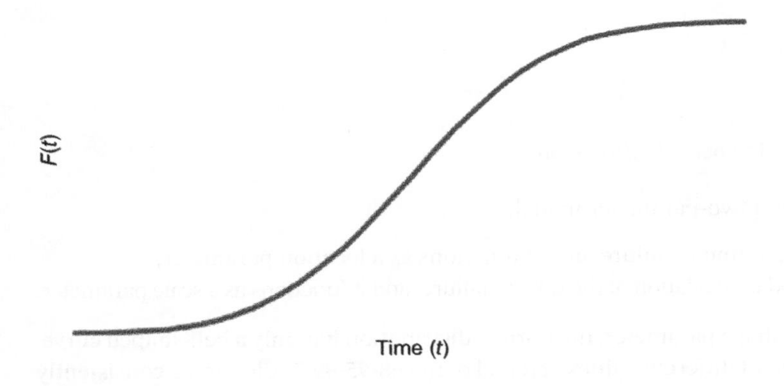

Figure 3.24 Normal cumulative distribution function.

The normal reliability function can be written as

$$R(t)=1-F(t)=1-\int_0^t \frac{1}{\sigma\sqrt{2\pi}} e^{-\frac{1}{2}\left(\frac{t-\mu}{\sigma}\right)^2} dt=\int_t^\infty \frac{1}{\sigma\sqrt{2\pi}} e^{-\frac{1}{2}\left(\frac{t-\mu}{\sigma}\right)^2} dt \tag{3.36}$$

Figure 3.25 illustrates examples of the normal reliability function.

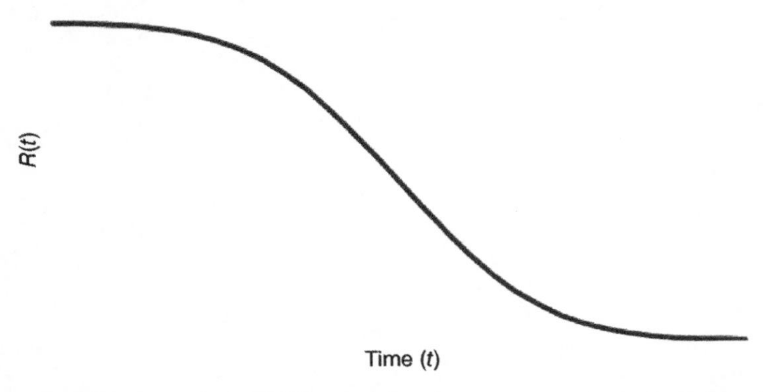

Figure 3.25 Normal reliability function.

The normal failure rate function can be written as

$$h(t)=\frac{f(t)}{R(t)}=\frac{\frac{1}{\sigma\sqrt{2\pi}}e^{-\frac{1}{2}\left(\frac{t-\mu}{\sigma}\right)^2}}{\int_t^{\infty}\frac{1}{\sigma\sqrt{2\pi}}e^{-\frac{1}{2}\left(\frac{t-\mu}{\sigma}\right)^2}dt} \tag{3.37}$$

Figure 3.26 illustrates examples of the normal failure rate function. There is an increasing pattern of the failure rate over time.

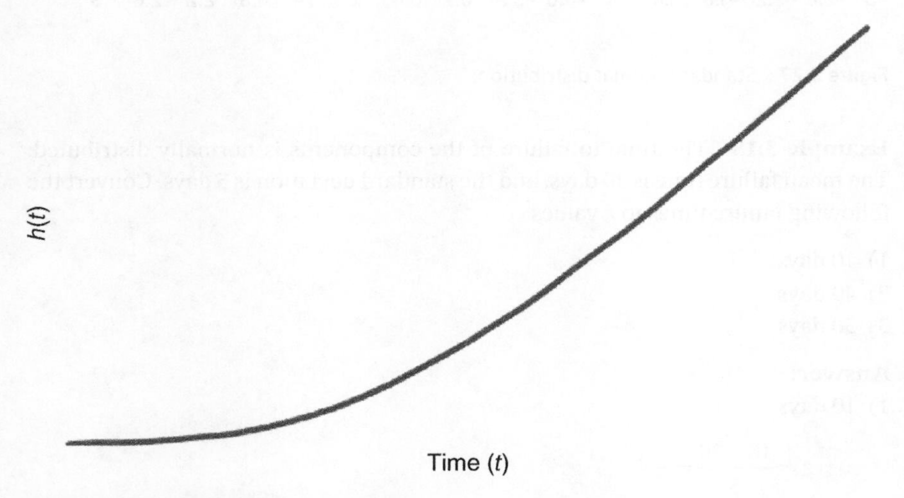

Figure 3.26 Normal failure rate function.

Since the equations of the normal distribution are complex, the standard normal distribution can be used as an alternative. A z-score can be calculated to convert an original normal distribution into the standard normal distribution.

$$z=\frac{t-\mu}{\sigma} \tag{3.38}$$

Figure 3.27 shows an example of the standard normal distribution. The mean is zero and the standard deviation is 1 in the standard normal distribution.

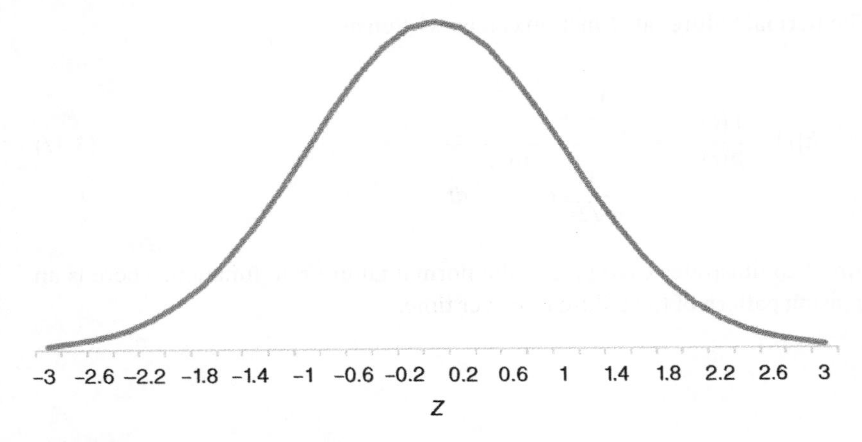

Figure 3.27 Standard normal distribution.

Example 3.18 The time to failure of the components is normally distributed. The mean failure time is 30 days, and the standard deviation is 5 days. Convert the following failure times to z values.

1) 10 days
2) 40 days
3) 30 days

Answer:

1) 10 days

$$z = \frac{10-30}{5} = -4$$

2) 40 days

$$z = \frac{40-30}{5} = 2$$

3) 30 days

$$z = \frac{30-30}{5} = 0$$

The standard normal table can be used to efficiently find the values of the cumulative distribution function of the normal distribution (Appendix B).

Example 3.19 Find the cumulative distribution function value of $z = 1.43$ in a standard normal distribution.

Answer:

$$\Phi(1.43) = 0.92364$$

Example 3.20 The time to failure of components is normally distributed. The mean failure time is 30 days, and the standard deviation is 5 days.

1) Calculate the probability of failure by 37 days.
2) Calculate the probability of survival after 42 days.

Answer:

1) Calculate the probability of failure by 37 days.

$$z = \frac{37-30}{5} = 1.4$$

$$\Phi(1.4) = 0.91924$$

2) Calculate the probability of survival after 42 days.

$$z = \frac{42-30}{5} = 2.4$$

$$\Phi(2.4) = 0.99180$$

$$1 - 0.99180 = 0.0082$$

The time to failure can be calculated using the standard normal distribution.

$$t = \mu + z\sigma \tag{3.39}$$

Example 3.21 The time to failure of components is normally distributed. The mean failure time is 30 days, and the standard deviation is 5 days. Calculate the failure time corresponding to a z value of 1.96.

Answer:

$$t = \mu + z\sigma = 30 + 1.96 \times 5 = 39.8 \text{ days}$$

Example 3.22 The time to failure of components is normally distributed. The mean failure time is 30 days, and the standard deviation is 5 days. A company wants to know when 5% of the components will survive.

Answer:

$$\Phi^{-1}(0.95) = 1.645$$

$$t = \mu + z\sigma = 30 + 1.645 \times 5 = 38.225 \text{ day}$$

Example 3.23 Two parameters of the normal distribution are not directly provided. Five percent of components will fail by 200 days, and another 5% of components will operate after 400 days. Calculate the reliability at 250 days.

Answer:

$$F(400) = \Phi\left(\frac{400-\mu}{\sigma}\right) = 0.95$$

$$\Phi^{-1}(0.95) = 1.645$$

$$\frac{400-\mu}{\sigma} = 1.645$$

$$F(200) = \Phi\left(\frac{200-\mu}{\sigma}\right) = 0.05$$

$$\Phi^{-1}(0.05) = -1.645$$

$$\frac{200-\mu}{\sigma} = -1.645$$

$$\frac{400-\mu}{\sigma} = 1.645 => 1.645\sigma = 400-\mu$$

$$\frac{200-\mu}{\sigma} = -1.645 => -1.645\sigma = 200-\mu$$

$$\mu = 300,\ \sigma = 60.79$$

$$R(250) = 1-\Phi\left(\frac{250-300}{60.79}\right) = 1-\Phi(-0.823) = 1-0.2061 = 0.7939$$

3.6.1 Excel Practice

The Excel functions can be used to efficiently calculate the normal distribution.

- NORMDIST(t, μ, σ, TRUE)
  - TRUE = cumulative distribution function
  - FALSE = probability density function

The standard normal distribution can be calculated using the Excel function.

- NORMSDIST(z)

The z value can be determined using the Excel function.

- NORMSINV($F(t)$)

3.6.2 Minitab Practice

The PDF with varying parameter values of the normal distribution can be constructed using Minitab. Let's assume that we want to compare three different sigma values $\sigma = 5, 10, 15$ with $\mu = 30$ hours.

The setup of the PDF of the normal distribution is shown in Figure 3.28.

Probability Distribution Plot: Vary Parameters

Distribution: Normal

Means: 30

Standard deviations: 5 10 15

Multiple Graphs...

Help | OK | Cancel

Figure 3.28 Normal distribution PDF setup.

The PDF of the normal distribution with multiple sigma values can be constructed (Figure 3.29).

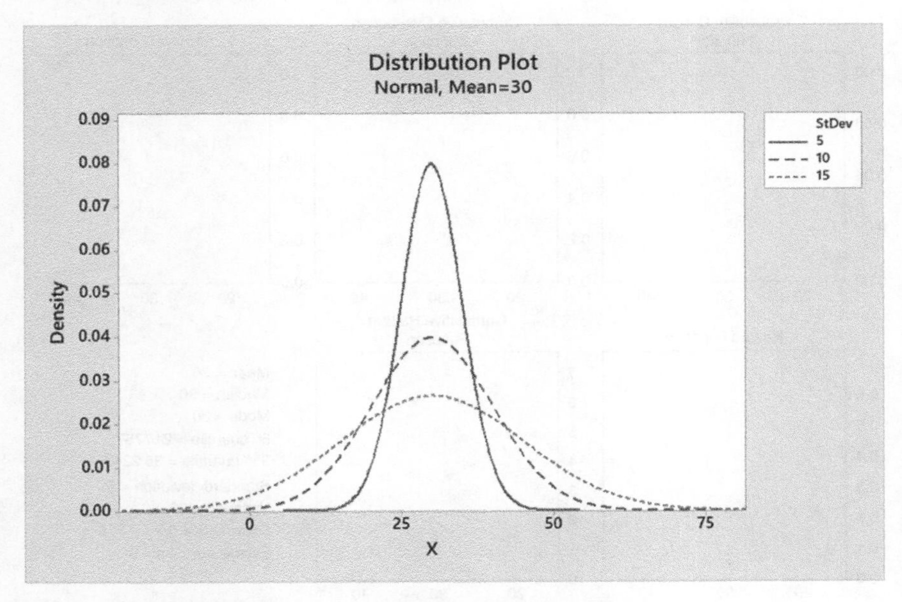

Figure 3.29 Normal distribution PDF with varying sigma values.

3.6.3 Python Practice

If we know the parameter information of the normal distribution, PDF, CDF, $R(t)$, $h(t)$, and cumulative $h(t)$ plots can be constructed using Python. Let's assume that $\mu = 30$ hours and $\sigma = 5$.

The Python codes used to construct the normal distribution plots are shown in Figure 3.30.

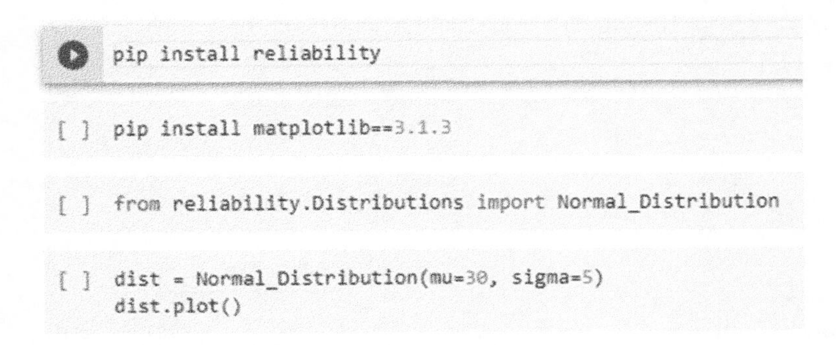

pip install reliability
[ ] pip install matplotlib==3.1.3
[ ] from reliability.Distributions import Normal_Distribution
[ ] dist = Normal_Distribution(mu=30, sigma=5)
dist.plot()

Figure 3.30 Python codes used to construct the normal distribution plots.

After running all the codes, the plots can be created as shown in Figure 3.31.

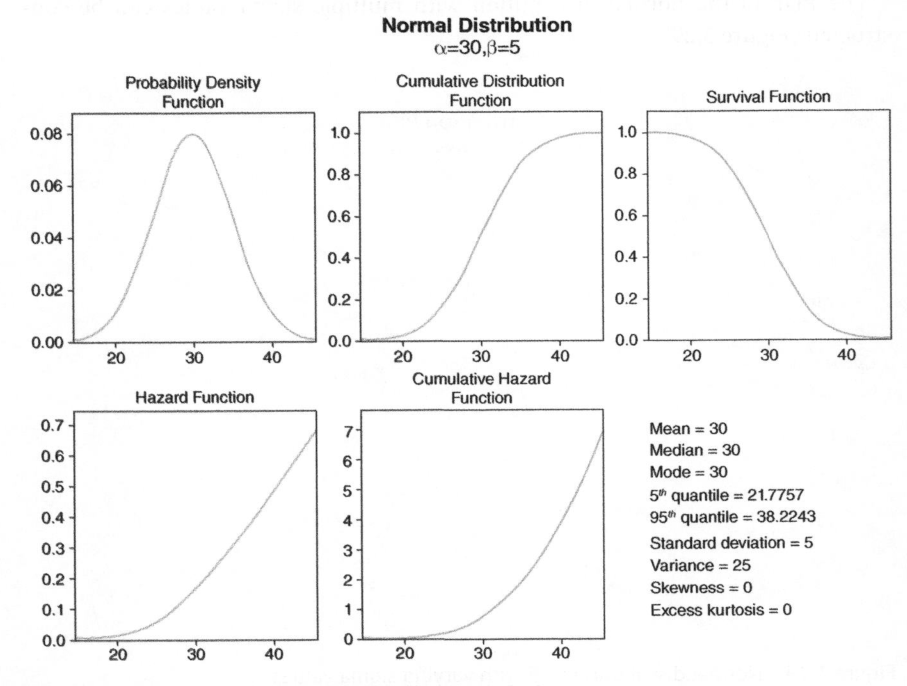

Figure 3.31 The normal distribution plots with Python.

3.7 Lognormal Distribution

The lognormal distribution has an asymmetric shape. The curve is skewed to the right side, called positively skewed. The lognormal distribution has flexible shapes, so it can be a good companion to the Weibull distribution. The lognormal distribution is often applied to the time to repair components or time to fatigue failures.

The lognormal distribution is closely related to the normal distribution. If the random variable, t, shows a lognormal distribution, $\ln(t)$ will reveal a normal distribution.

The equation of the lognormal probability density function can be written as

$$f(t) = \frac{1}{\sigma t\sqrt{2\pi}} e^{-\left(\frac{1}{2\sigma^2}\right)(\ln t - \ln T_{50})^2} \tag{3.40}$$

The two-parameter model is used.

- T_{50} is a median lifetime and it functions as a location parameter.
- σ functions as a shape parameter.

Figure 3.32 illustrates the lognormal probability density function.

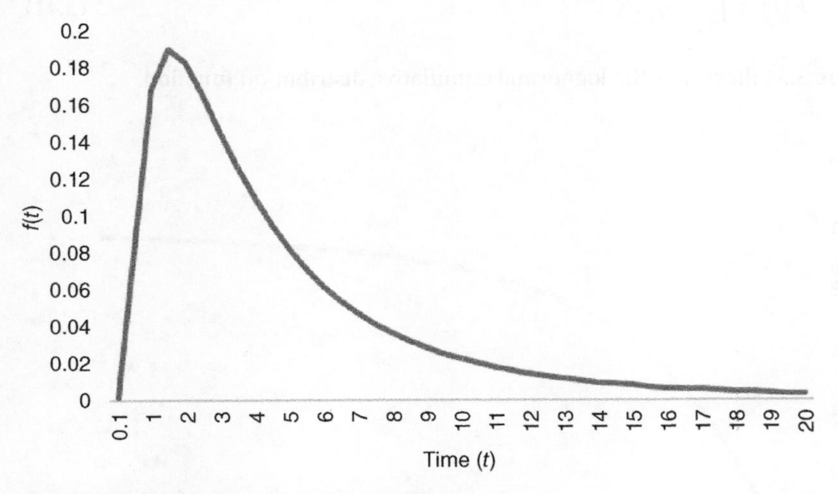

Figure 3.32 Lognormal probability density function.

As seen in Figure 3.33, the lognormal distribution takes a variety of shapes. For $\sigma = 0.2$, it is close to the shape of the normal distribution.

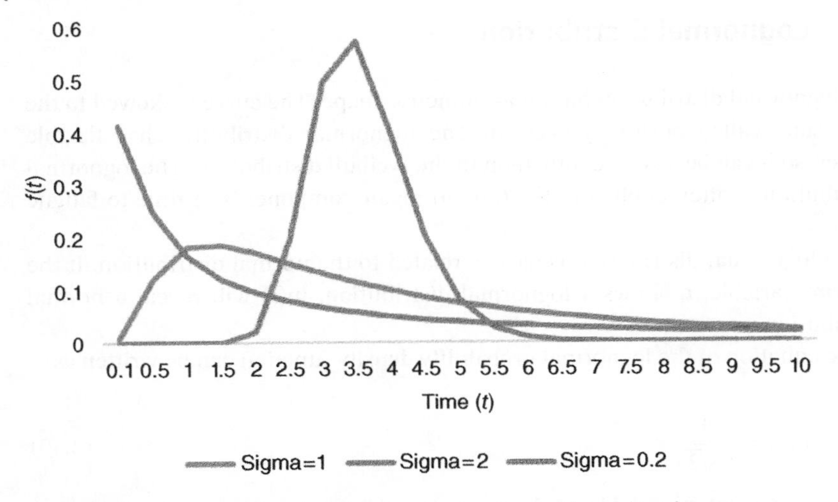

Figure 3.33 Lognormal probability density function with varying shapes.

The equation of the lognormal cumulative distribution function can be written as

$$F(t) = \int_0^t \frac{1}{\sigma t\sqrt{2\pi}} e^{-\left(\frac{1}{2\sigma^2}\right)^{(\ln t - \ln T_{50})^2}} \tag{3.41}$$

Figure 3.34 illustrates the lognormal cumulative distribution function.

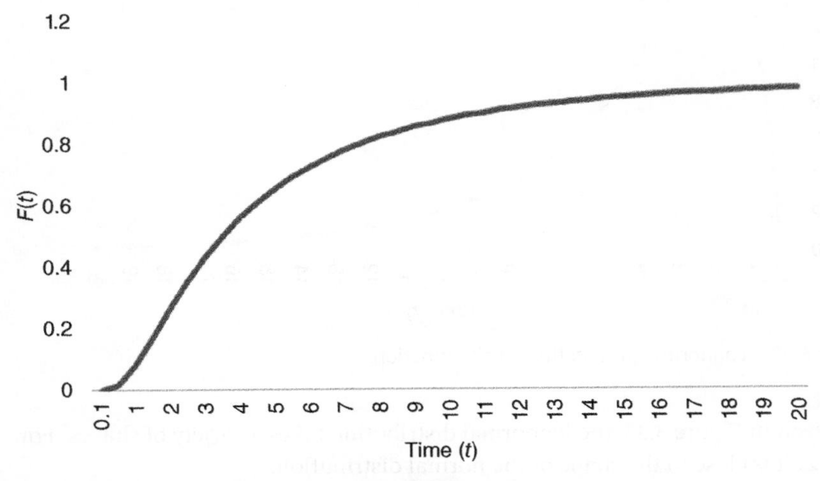

Figure 3.34 Lognormal cumulative distribution function.

The equations of the lognormal distribution are complex. The relationship between the lognormal distribution and the normal distribution can be used for efficient calculations.

$$F(t)=\Phi(z)=\Phi\left(\frac{t-\mu}{\sigma}\right)=\Phi\left(\frac{\ln(t_a)-\ln(T_{50})}{\sigma}\right)=\Phi\left(\frac{\ln\left(\frac{t_a}{T_{50}}\right)}{\sigma}\right)=\Phi\left(\frac{\ln t_a-\mu)}{\sigma}\right) \quad (3.42)$$

- t_a is a time to failure that shows a lognormal distribution.
- t is a time to failure that shows a normal distribution.

Example 3.24 Components' failure characteristics are related to the lognormal distribution with $T_{50} = 100$ days, and $\sigma = 5$.

1) What percentage of failures is expected after 120 days?
2) What is the reliability at 130 days?

Answer:

1) What percentage of failures is expected after 120 days?

$$F(120)=\Phi\left(\frac{\ln\left(\frac{120}{100}\right)}{5}\right)=\Phi(0.036)=0.484$$

We expect 48.4% of failures after 120 days.

2) What is the reliability at 130 days?

$$R(130)=1-F(130)=1-\Phi\left(\frac{\ln\left(\frac{130}{100}\right)}{5}\right)=1-\Phi(0.052)=1-0.4801=0.5199$$

The lognormal failure rate function has a variety of shapes depending on the σ values.

- For $\sigma \geq 2$, a high failure rate is shown at the beginning, and it is decreasing over time.
- For $\sigma \leq 0.5$, the failure rate is increasing over time, such as with wearout failure characteristics.

The equation of the mean time to failure (MTTF) of the lognormal distribution is

$$\text{MTTF} = T_{50}e^{\left(\frac{\sigma^2}{2}\right)} \quad (3.43)$$

Example 3.25 Components' failure characteristics are related to the lognormal distribution with $T_{50} = 100$ days, and $\sigma = 1$. Calculate the mean time to failure.

Answer:

$$\text{MTTF} = T_{50}e^{\left(\frac{\sigma^2}{2}\right)} = 100e^{\left(\frac{5^2}{2}\right)} = 165 \text{ days}$$

The median time (T_{50}) can be calculated based on the cumulative distribution function.

$$T_{50} = t_a e^{-\sigma z} = t_a e^{-\sigma\Phi^{-1}(\text{CDF})} \tag{3.44}$$

The shape parameter (σ) can be calculated:

$$\sigma = \frac{\ln\left(\frac{t_a}{T_{50}}\right)}{z} = \frac{\ln\left(\frac{t_a}{T_{50}}\right)}{\Phi^{-1}(\text{CDF})} \tag{3.45}$$

The failure time can be calculated:

$$t_a = T_{50}e^{\sigma z} = T_{50}e^{\sigma\Phi^{-1}(\text{CDF})} \tag{3.46}$$

Example 3.26 Five percent of components fail by 50,000 hours. The failure characteristic is related to the lognormal distribution, and $\sigma = 0.5$. Calculate T_{50}.

Answer:

$$T_{50} = 50{,}000e^{-0.5\Phi^{-1}(0.05)} = 50{,}000e^{-0.5(-1.645)} = 113{,}809 \text{ hours}$$

Example 3.27 The failure of components is modeled by the lognormal distribution with $\sigma = 0.5$ and $T_{50} = 40{,}000$ hours. How many hours does it take to reach 10% cumulative failures?

Answer:

$$t_a = 40{,}000e^{0.5\Phi^{-1}(0.1)} = 40{,}000e^{0.5(-1.28)} = 21{,}092 \text{ hours}$$

3.7.1 Excel Practice

The Excel functions can be used to efficiently calculate the lognormal distribution.

- LOGNORMDIST(t_a, μ, σ, TRUE) or LOGNORMDIST(t_a, $\ln(T_{50})$, σ, TRUE)
- TRUE = cumulative distribution function
- FALSE = probability density function

3.7.2 Minitab Practice

The PDF with varying parameter values of the lognormal distribution can be constructed using Minitab. Let's assume that we want to compare three different sigma values $\sigma = 0.5, 1, 1.5$ with $\mu = 1.8$ hours. Figure 3.35 shows the setup of the PDF of the lognormal distribution.

Probability Distribution Plot: Vary Parameters

Distribution:
Lognormal

Locations: 1.8
Scales: 0.5 1 1.5
Thresholds: 0.0

Multiple Graphs...

Help OK Cancel

Figure 3.35 Lognormal PDF setup.

The PDF of the lognormal distribution with varying sigma values can be constructed as seen in Figure 3.36.

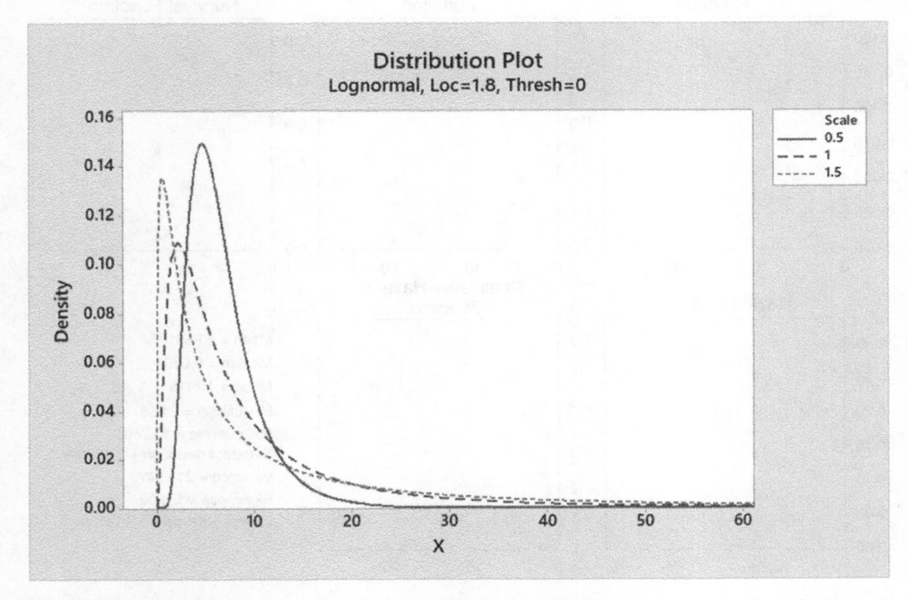

Figure 3.36 Lognormal PDF with varying sigma values using Minitab.

3.7.3 Python Practice

If we know the parameter information of the lognormal distribution, PDF, CDF, $R(t)$, $h(t)$, and cumulative $h(t)$ plots can be constructed using Python. Let's assume that $\mu = 1.8$ hours and $\sigma = 0.5$. Figure 3.37 shows the Python codes used to construct the lognormal distribution plots.

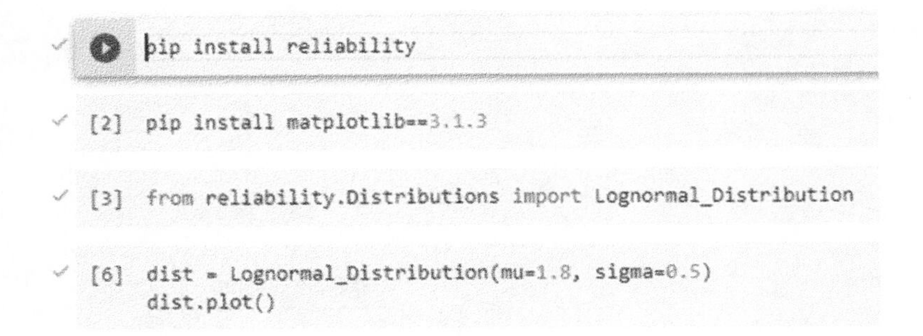

pip install reliability
[2] pip install matplotlib==3.1.3
[3] from reliability.Distributions import Lognormal_Distribution
[6] dist = Lognormal_Distribution(mu=1.8, sigma=0.5)
dist.plot()

Figure 3.37 Python codes used to construct the lognormal distribution plots.

After running all the codes, the plots can be created as shown in Figure 3.38.

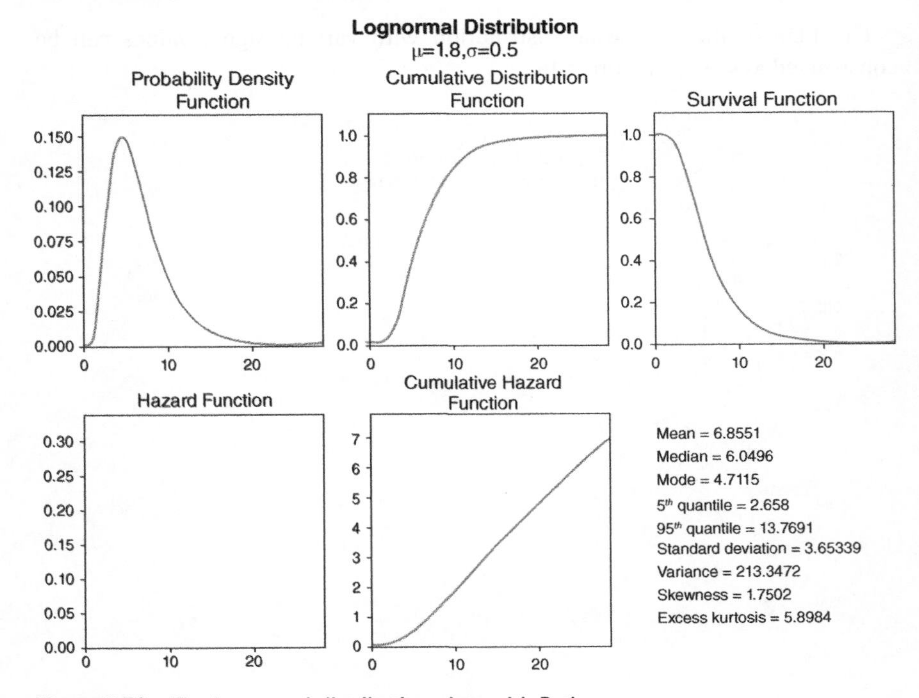

Figure 3.38 The lognormal distribution plots with Python.

Tip! Distribution Explorer

The distribution explorer can be constructed as an interactive tool to explore different parameter values of probability distributions. Figure 3.39 gives the Python codes to run the distribution explorer.

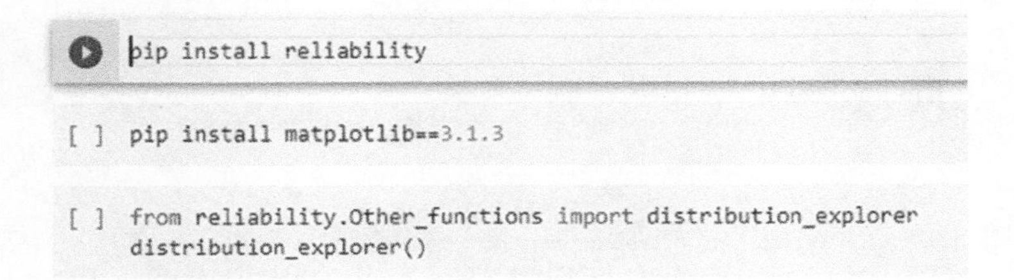
```
pip install reliability

[ ] pip install matplotlib==3.1.3

[ ] from reliability.Other_functions import distribution_explorer
    distribution_explorer()
```

Figure 3.39 Python codes for the distribution explorer.

After running all codes, the distribution explorer GUI can be created (Figure 3.40). There are several options to select the particular life distribution. If we drag the parameter, the values will be changed, and the shape of the charts will be updated instantly. Since Google Colab does not provide support to emulate GUI, the Python IDLE platform (https://docs.python.org/3/library/idle.html) could be used to run the GUI.

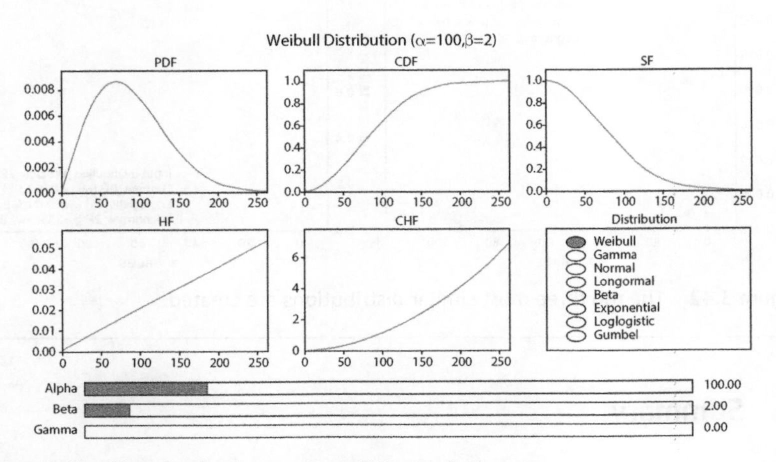

Figure 3.40 The distribution explorer GUI with Python.

Tip! Similar Distribution Finder

Based on the Python codes, we could find the top three most similar distributions to the particular life distribution with certain parameter values. Figure 3.41 shows the Python codes to generate similar distributions. This example uses a Weibull distribution with $\alpha = 30$, $\beta = 1.5$.

```
pip install reliability
```

```
[ ] pip install matplotlib==3.1.3
```

```
[ ] from reliability.Distributions import Weibull_Distribution
    from reliability.Other_functions import similar_distributions
```

```
[ ] dist = Weibull_Distribution(alpha=30,beta=1.5)
    similar_distributions(distribution=dist,include_location_shifted=False)
```

Figure 3.41 The Python codes to generate similar distributions.

After running all codes, the top three similar distributions with parameter values are provided (Figure 3.42).

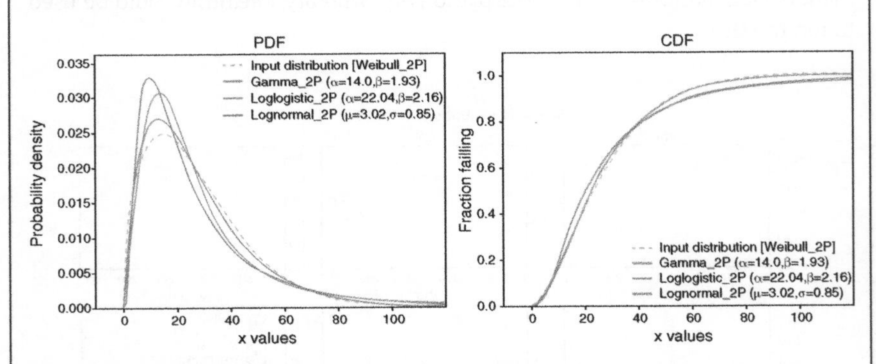

Figure 3.42 The top three most similar distributions are created.

3.8 Summary

- The random variable can be categorized into discrete and continuous variables.
- A discrete probability distribution accounts for discrete random variables that have probability values at discrete points.
- A probability mass function (PMF) shows the magnitude of the probability of each discrete point.

- A cumulative distribution function (CDF) is the probability that a random variable (X) will take a value less than or equal to a certain value (x).
- A continuous probability distribution shows the probabilities that are areas under the curve (integrals).
- The probability density function is a derivative of the cumulative distribution function.
- The cumulative distribution function provides the probability of the random variable at a value less than or equal to a certain value.
- A failure rate (or hazard rate) is the number of failures per unit of time.
- The unique feature of the exponential distribution is that it has a constant failure rate function.
- The Weibull distribution is versatile, so it can be applied to increasing, constant, and decreasing failure rates over time.
- The normal distribution can be used to model the items that reveal fatigue or wearout failures.
- The lognormal distribution has flexible shapes, so it can be a good companion to the Weibull distribution.

Exercises

1 Describe the difference between discrete and continuous random variables.

2 Explain the probability mass function (PMF) and cumulative distribution function (CDF) of a discrete random variable.

3 The service life of a mobile phone battery has a Weibull distribution with $\alpha = 2$ years and $\beta = 1.5$. Calculate the probability that a battery will last after 2.5 years.

4 The wearout failures of a dump truck's tires are characterized by the Weibull distribution with $\alpha = 20{,}000$ km and $\beta = 2.5$. Estimate the distance at which 10% of the tires will fail.

5 Diodes exhibit a constant failure rate of 0.2%/K. Calculate the probability that the diodes will last after 15,000 hours of use.

6 The components in a vacuum machine show failure characteristics according to the Weibull distribution with $\alpha = 2{,}000$ hours and $\beta = 2$. Calculate the probability that a component that has been operating for 1000 hours survives an additional 500 hours.

7 The life of a tire tread is normally distributed with a mean of 24 months and a standard deviation of 7 months. Calculate the probability of failure by 30 months.

8 The life of a hearing aid battery is normally distributed, with a mean of 2 years and a standard deviation of 1.5 years. The hearing aid company wants to know when 10% of batteries will fail.

9 Fatigue wearout characteristics of a spring reveal a lognormal distribution with $T_{50} = 3000$ hours and $\sigma = 0.2$. Calculate the mean time to failure. Calculate the design life for a reliability of 0.95.

10 A semiconductor experiences constant failure rates with a MTTF of 1300 hours. Calculate the median time to failure. Calculate the design life for 0.90 reliability.

Appendix A

Table for Gamma Function.

α	$\Gamma(\alpha)$	α	$\Gamma(\alpha)$	α	$\Gamma(\alpha)$	α	$\Gamma(\alpha)$	α	$\Gamma(\alpha)$
1.00	1.000000	1.20	0.918169	1.40	0.887264	1.60	0.893515	1.80	0.931384
1.02	0.988844	1.22	0.913106	1.42	0.886356	1.62	0.895924	1.82	0.936845
1.04	0.978438	1.24	0.908521	1.44	0.885805	1.64	0.898642	1.84	0.942612
1.03	0.968744	1.26	0.904397	1.46	0.885604	1.66	0.901668	1.86	0.948687
1.08	0.959725	1.28	0.900718	1.48	0.885747	1.68	0.905001	1.88	0.955071
1.10	0.951351	1.30	0.897471	1.50	0.886227	1.70	0.908639	1.90	0.961766
1.12	0.943590	1.32	0.894640	1.52	0.887039	1.72	0.912581	1.92	0.968774
1.14	0.936416	1.34	0.892216	1.54	0.888178	1.74	0.916826	1.94	0.976099
1.16	0.929803	1.36	0.890185	1.56	0.889639	1.76	0.921375	1.96	0.983743
1.18	0.923728	1.38	0.888537	1.58	0.891420	1.78	0.926227	1.98	0.991708
1.20	0.918169	1.40	0.887264	1.60	0.893515	1.80	0.931384	2.00	1.000000

Appendix B

Standard Normal Table.

z	0.00	0.01	0.02	0.03	0.04	0.05	0.06	0.07	0.08	0.09
−3.4	0.0003	0.0003	0.0003	0.0003	0.0003	0.0003	0.0003	0.0003	0.0003	0.0002
−3.3	0.0005	0.0005	0.0005	0.0004	0.0004	0.0004	0.0004	0.0004	0.0004	0.0003
−3.2	0.0007	0.0007	0.0006	0.0006	0.0006	0.0006	0.0006	0.0005	0.0005	0.0005
−3.1	0.0010	0.0009	0.0009	0.0009	0.0008	0.0008	0.0008	0.0008	0.0007	0.0007
−3.0	0.0013	0.0013	0.0013	0.0012	0.0012	0.0011	0.0011	0.0011	0.0010	0.0010

(Continued)

(Continued)

−2.9	0.0019	0.0018	0.0018	0.0017	0.0016	0.0016	0.0015	0.0015	0.0014	0.0014
−2.8	0.0026	0.0025	0.0024	0.0023	0.0023	0.0022	0.0021	0.0021	0.0020	0.0019
−2.7	0.0035	0.0034	0.0033	0.0032	0.0031	0.0030	0.0029	0.0028	0.0027	0.0026
−2.6	0.0047	0.0045	0.0044	0.0043	0.0041	0.0040	0.0039	0.0038	0.0037	0.0036
−2.5	0.0062	0.0060	0.0059	0.0057	0.0055	0.0054	0.0052	0.0051	0.0049	0.0048
−2.4	0.0082	0.0080	0.0078	0.0075	0.0073	0.0071	0.0069	0.0068	0.0066	0.0064
−2.3	0.0107	0.0104	0.0102	0.0099	0.0096	0.0094	0.0091	0.0089	0.0087	0.0084
−2.2	0.0139	0.0136	0.0132	0.0129	0.0125	0.0122	0.0119	0.0116	0.0113	0.0110
−2.1	0.0179	0.0174	0.0170	0.0166	0.0162	0.0158	0.0154	0.0150	0.0146	0.0143
−2.0	0.0228	0.0222	0.0217	0.0212	0.0207	0.0202	0.0197	0.0192	0.0188	0.0183
−1.9	0.0287	0.0281	0.0274	0.0268	0.0262	0.0256	0.0250	0.0244	0.0239	0.0233
−1.8	0.0359	0.0351	0.0344	0.0336	0.0329	0.0322	0.0314	0.0307	0.0301	0.0294
−1.7	0.0446	0.0436	0.0427	0.0418	0.0409	0.0401	0.0392	0.0384	0.0375	0.0367
−1.6	0.0548	0.0537	0.0526	0.0516	0.0505	0.0495	0.0485	0.0475	0.0465	0.0455
−1.5	0.0668	0.0655	0.0643	0.0630	0.0618	0.0606	0.0594	0.0582	0.0571	0.0559
−1.4	0.0808	0.0793	0.0778	0.0764	0.0749	0.0735	0.0721	0.0708	0.0694	0.0681
−1.3	0.0968	0.0951	0.0934	0.0918	0.0901	0.0885	0.0869	0.0853	0.0838	0.0823
−1.2	0.1151	0.1131	0.1112	0.1093	0.1075	0.1056	0.1038	0.1020	0.1003	0.0985
−1.1	0.1357	0.1335	0.1314	0.1292	0.1271	0.1251	0.1230	0.1210	0.1190	0.1170
−1.0	0.1587	0.1562	0.1539	0.1515	0.1492	0.1469	0.1446	0.1423	0.1401	0.1379
−0.9	0.1841	0.1814	0.1788	0.1762	0.1736	0.1711	0.1685	0.1660	0.1635	0.1611
−0.8	0.2119	0.2090	0.2061	0.2033	0.2005	0.1977	0.1949	0.1922	0.1894	0.1867
−0.7	0.2420	0.2389	0.2358	0.2327	0.2296	0.2266	0.2236	0.2206	0.2177	0.2148
−0.6	0.2743	0.2709	0.2676	0.2643	0.2611	0.2578	0.2546	0.2514	0.2483	0.2451
−0.5	0.3085	0.3050	0.3015	0.2981	0.2946	0.2912	0.2877	0.2843	0.2810	0.2776
−0.4	0.3446	0.3409	0.3372	0.3336	0.3300	0.3264	0.3228	0.3192	0.3156	0.3121
−0.3	0.3821	0.3783	0.3745	0.3707	0.3669	0.3632	0.3594	0.3557	0.3520	0.3483
−0.2	0.4207	0.4168	0.4129	0.4090	0.4052	0.4013	0.3974	0.3936	0.3897	0.3859
−0.1	0.4602	0.4562	0.4522	0.4483	0.4443	0.4404	0.4364	0.4325	0.4286	0.4247
0.0	0.5000	0.4960	0.4920	0.4880	0.4840	0.4801	0.4761	0.4721	0.4681	0.4641

z	0.00	0.01	0.02	0.03	0.04	0.05	0.06	0.07	0.08	0.09
0.0	0.5000	0.5040	0.5080	0.5120	0.5160	0.5199	0.5239	0.5279	0.5319	0.5359
0.1	0.5398	0.5438	0.5478	0.5517	0.5557	0.5596	0.5636	0.5675	0.5714	0.5753
0.2	0.5793	0.5832	0.5871	0.5910	0.5948	0.5987	0.6026	0.6064	0.6103	0.6141
0.3	0.6179	0.6217	0.6255	0.6293	0.6331	0.6368	0.6406	0.6443	0.6480	0.6517
0.4	0.6554	0.6591	0.6628	0.6664	0.6700	0.6736	0.6772	0.6808	0.6844	0.6879
0.5	0.6915	0.6950	0.6985	0.7019	0.7054	0.7088	0.7123	0.7157	0.7190	0.7224
0.6	0.7257	0.7291	0.7324	0.7357	0.7389	0.7422	0.7454	0.7486	0.7517	0.7549
0.7	0.7580	0.7611	0.7642	0.7673	0.7704	0.7734	0.7764	0.7794	0.7823	0.7852
0.8	0.7881	0.7910	0.7939	0.7967	0.7995	0.8023	0.8051	0.8078	0.8106	0.8133
0.9	0.8159	0.8186	0.8212	0.8238	0.8264	0.8289	0.8315	0.8340	0.8365	0.8389
1.0	0.8413	0.8438	0.8461	0.8485	0.8508	0.8531	0.8554	0.8577	0.8599	0.8621
1.1	0.8643	0.8665	0.8686	0.8708	0.8729	0.8749	0.8770	0.8790	0.8810	0.8830
1.2	0.8849	0.8869	0.8888	0.8907	0.8925	0.8944	0.8962	0.8980	0.8997	0.9015
1.3	0.9032	0.9049	0.9066	0.9082	0.9099	0.9115	0.9131	0.9147	0.9162	0.9177
1.4	0.9192	0.9207	0.9222	0.9236	0.9251	0.9265	0.9279	0.9292	0.9306	0.9319
1.5	0.9332	0.9345	0.9357	0.9370	0.9382	0.9394	0.9406	0.9418	0.9429	0.9441
1.6	0.9452	0.9463	0.9474	0.9484	0.9495	0.9505	0.9515	0.9525	0.9535	0.9545
1.7	0.9554	0.9564	0.9573	0.9582	0.9591	0.9599	0.9608	0.9616	0.9625	0.9633
1.8	0.9641	0.9649	0.9656	0.9664	0.9671	0.9678	0.9686	0.9693	0.9699	0.9706
1.9	0.9713	0.9719	0.9726	0.9732	0.9738	0.9744	0.9750	0.9756	0.9761	0.9767
2.0	0.9772	0.9778	0.9783	0.9788	0.9793	0.9798	0.9803	0.9808	0.9812	0.9817
2.1	0.9821	09826	0.9830	0.9834	0.9838	0.9842	0.9846	0.9850	0.9854	0.9857
2.2	0.9861	0.9864	0.9868	0.9871	0.9875	0.9878	0.9881	0.9884	0.9887	0.9890
2.3	0.9893	0.9896	0.9898	0.9901	0.9904	0.9906	0.9909	0.9911	0.9913	0.9916
2.4	0.9918	0.9920	0.9922	0.9925	0.9927	0.9929	0.9931	0.9932	0.9934	0.9936
2.5	0.9938	0.9940	0.9941	0.9943	0.9945	0.9946	0.9948	0.9949	0.9951	0.9952
2.6	0.9953	0.9955	0.9956	0.9957	0.9959	0.9960	0.9961	0.9962	0.9963	0.9964
2.7	0.9965	0.9966	0.9967	0.9968	0.9969	0.9970	0.9971	0.9972	0.9973	0.9974
2.8	0.9974	0.9975	0.9976	0.9977	0.9977	0.9978	0.9979	0.9979	0.9980	0.9981
2.9	0.9981	0.9982	0.9982	0.9983	0.9984	0.9984	0.9985	0.9985	0.9986	0.9986

(Continued)

(Continued)

3.0	0.9987	0.9987	0.9987	0.9988	0.9988	0.9989	0.9989	0.9989	0.9990	0.9990
3.1	0.9990	0.9991	0.9991	0.9991	0.9992	0.9992	0.9992	0.9992	0.9993	0.9993
3.2	0.9993	0.9993	0.9994	0.9994	0.9994	0.9994	0.9994	0.9995	0.9995	0.9995
3.3	0.9995	0.9995	0.9995	0.9996	0.9996	0.9996	0.9996	0.9996	0.9996	0.9997
3.4	0.9997	0.9997	0.9997	0.9997	0.9997	0.9997	0.9997	0.9997	0.9997	0.9998

4

Reliability Data Plotting

Chapter Overview and Learning Objectives

- To understand the properties of the straight line.
- To learn the characteristics of the least squares fit and calculate the parameters of the straight line.
- To learn how to conduct linear rectification.
- To construct probability plots of various life distributions, including exponential, Weibull, normal, and lognormal distribution.
- To utilize Excel, Minitab, and Python to construct probability plots.

4.1 Straight Line Properties

In a previous chapter, we learn different parameters of various lifetime distributions, including exponential, Weibull, normal, and lognormal distributions. Graphical procedures could be useful to validate the assumptions of particular distributions to the data. Reliability data plotting also helps to estimate parameters of the distributions quickly.

Figure 4.1 illustrates the properties of a straight line. A straight line can be drawn based on two points. The slope parameter can be calculated as follows:

$$\beta_1 = \frac{y_2 - y_1}{x_2 - x_1} = \frac{\Delta y}{\Delta x} \tag{4.1}$$

The linear relationship can be shown through the equation

$$y = \beta_1 x \tag{4.2}$$

Reliability Analysis Using MINITAB and Python, First Edition. Jaejin Hwang.
© 2023 John Wiley & Sons, Inc. Published 2023 by John Wiley & Sons, Inc.
Companion Website: www.wiley.com\go\Hwang\ReliabilityAnalysisUsingMinitabandPython

The equation explains that the change in y is proportional to the change in x. The slope parameter, β_1, can also be considered a proportionality factor.

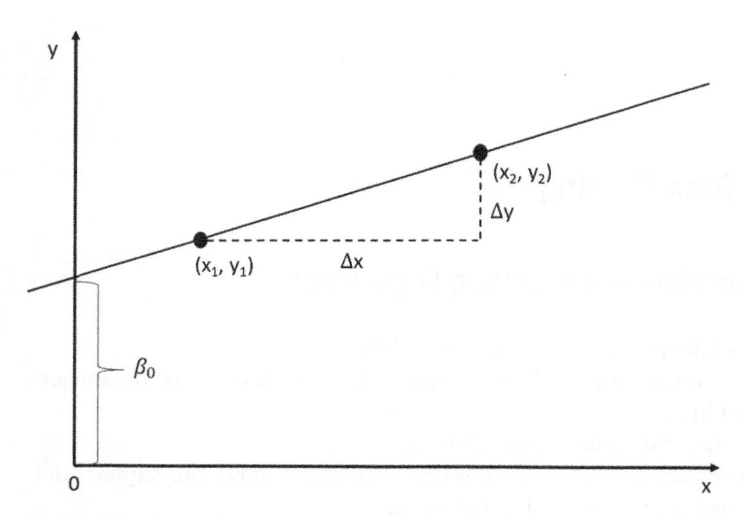

Figure 4.1 Properties of the straight line.

The intercept, β_0, can also be considered.

$$y = \beta_1 x + \beta_0 \tag{4.3}$$

At the point of $x = 0$, the β_0 can be easily obtained. The β_0 is the y coordinate point where the straight line crosses the y-axis, as seen in Figure 4.1.

A straight line can also be written in an alternative way:

$$\mathrm{A}x + \mathrm{B}y + \mathrm{C} = 0 \tag{4.4}$$

The A, B, and C are constants. For example, $4x–3y + 2 = 0$ describes a straight line.

Example 4.1 Find an equation for the straight line through two points, (1, 3) and (4, –2).

Answer:

$$\beta_1 = \frac{-2-3}{4-1} = -\frac{5}{3}$$

$$y = -\frac{5}{3}x + \beta_0$$

$$3 = -\frac{5}{3}(1) + \beta_0$$

$$\beta_0 = 4.67$$

$$y = -\frac{5}{3}x + 4.67$$

4.2 Least Squares Fit

When we collect the data through measurements, variation of the data can be expected. Figure 4.2 shows an example of the relationship between the number of failures and the temperature in Celsius. The scatter plot shows that the obtained data scatter around the expected straight line.

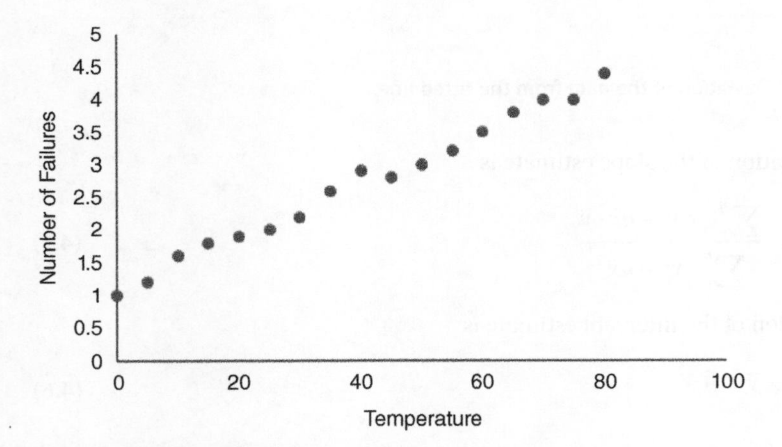

Figure 4.2 Example of the scatter plot.

The least squares fit or a regression analysis could be one way to establish the best-fit line for a given set of data. Figure 4.3 shows the deviation of one data point relative to the fitted line. The least squares fit determines the fitted line showing the minimum of the sum of the squares of the deviations from the fitted line.

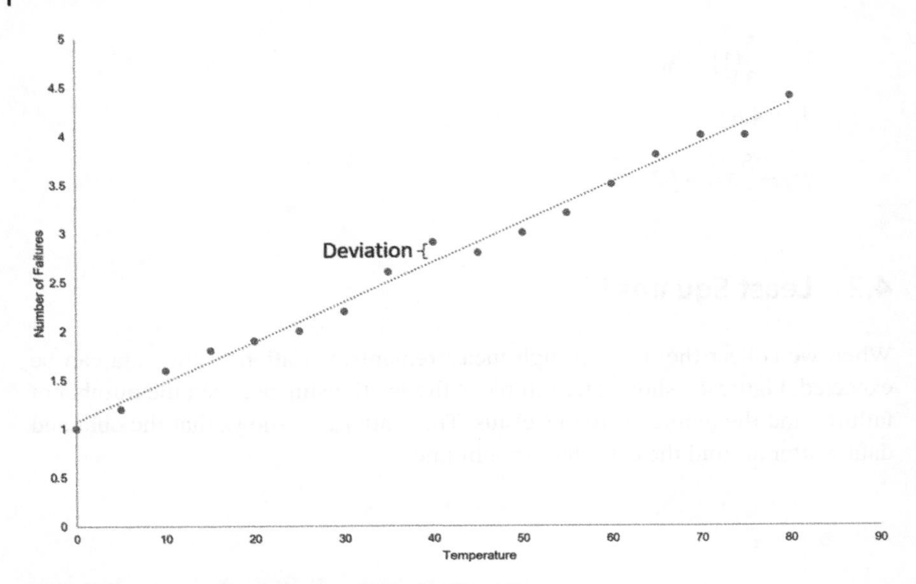

Figure 4.3 Deviation of the data from the fitted line.

The equation of the slope estimate is

$$\hat{\beta}_1 = \frac{\sum_{i=1}^{n} x_i y_i - n\bar{x} \cdot \bar{y}}{\sum_{i=1}^{n} x_i^2 - n\bar{x}^2} \tag{4.5}$$

The equation of the intercept estimate is

$$\hat{\beta}_0 = \bar{y} - \hat{\beta}_1 \bar{x} \tag{4.6}$$

Example 4.2 Construct the regression line formula based on the coordinate data of x and y (Table 4.1).

Table 4.1 The coordinate data of x and y.

x	y
1.1	3.0
2.0	4.1
3.2	5.2
4.1	6.0
5.2	7.3

Answer:

	x	*y*	x^2	*xy*
	1.1	3.0	1.2	3.3
	2.0	4.1	4.0	8.2
	3.2	5.2	10.2	16.6
	4.1	6.0	16.8	24.6
	5.2	7.3	27.0	38.0
Sums	15.6	25.6	59.3	90.7

$$\hat{\beta}_1 = \frac{\sum_{i=1}^{n} x_i y_i - n\bar{x}\cdot\bar{y}}{\sum_{i=1}^{n} x_i^2 - n\bar{x}^2} = \frac{90.7 - 5\times 3.12\times 5.12}{59.3 - 5\times 3.12^2} = 1.02$$

$$\hat{\beta}_0 = \bar{y} - \hat{\beta}_1\bar{x} = 5.12 - 1.02\times 3.12 = 1.94$$

$$\hat{y} = 1.02\text{x} + 1.94$$

4.2.1 Excel Practice

Parameters of the regression line can be computed using the Excel function:

- $\hat{\beta}_1$ = SLOPE(y data, x data)
- $\hat{\beta}_0$ = INTERCEPT(y data, x data)

The scatter plot can be used in Excel. As seen in Figure 4.4, the linear trend line can be added. By checking [Display Equation on chart], the linear regression formula can be shown in the chart.

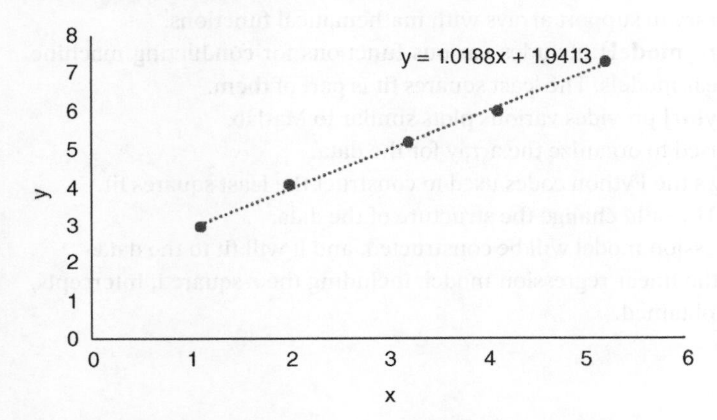

Figure 4.4 A scatter plot with the regression line.

4.2.2 Minitab Practice

Minitab software can be used to construct a regression line.

- The data table can be initially set. Use *x* and *y* data from Example 4.2.
- Stat -> Regression -> Fitted Line Plot
- Select [Response] as *y* data, and [Predictor] as *x* data. Hit [OK].
- The chart will be created as seen in Figure 4.5.

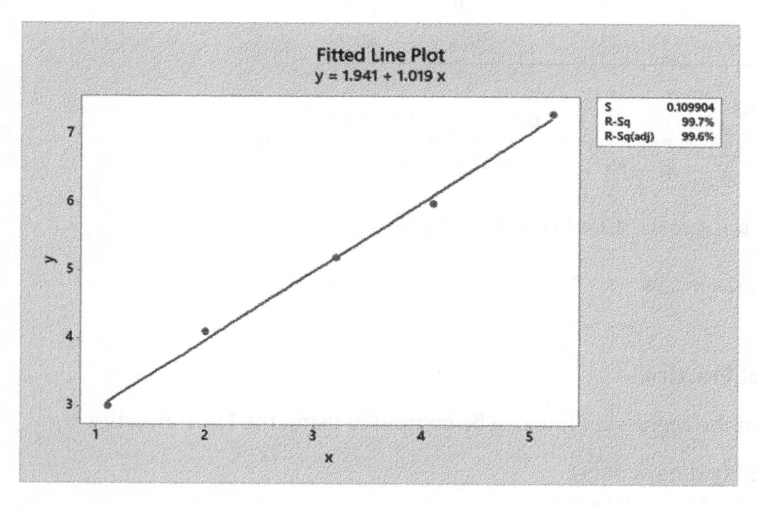

Figure 4.5 A scatter plot with the regression line using Minitab.

4.2.3 Python Practice

[numpy] is a library to support arrays with mathematical functions.

[sklearn.linear_model] includes various functions for conducting machine learning with linear models. The least squares fit is part of them.

[matplotlib.pylot] provides various plots similar to Matlab.

[np.array] is used to organize the array for the data.

Figure 4.6 shows the Python codes used to construct the least squares fit.

[reshape(–1,1)] would change the structure of the data.

The linear regression model will be constructed, and it will fit to the data.

Parameters of the linear regression model, including the *r*-squared, intercepts, and slope can be obtained.

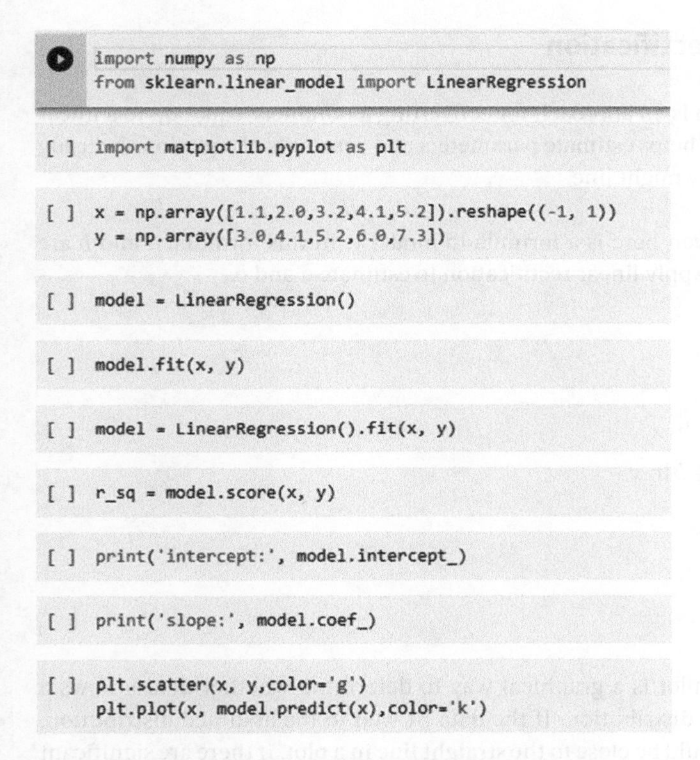

```
import numpy as np
from sklearn.linear_model import LinearRegression

[ ] import matplotlib.pyplot as plt

[ ] x = np.array([1.1,2.0,3.2,4.1,5.2]).reshape((-1, 1))
    y = np.array([3.0,4.1,5.2,6.0,7.3])

[ ] model = LinearRegression()

[ ] model.fit(x, y)

[ ] model = LinearRegression().fit(x, y)

[ ] r_sq = model.score(x, y)

[ ] print('intercept:', model.intercept_)

[ ] print('slope:', model.coef_)

[ ] plt.scatter(x, y,color='g')
    plt.plot(x, model.predict(x),color='k')
```

Figure 4.6 Python codes for creating the least squares fit.

After running all the codes, the least squares fit can be created on the chart (Figure 4.7).

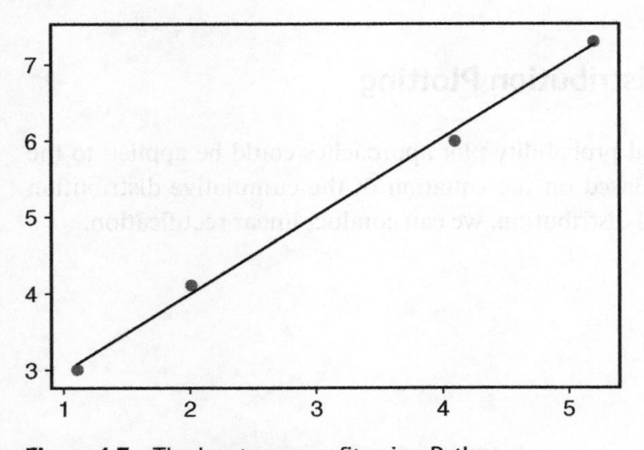

Figure 4.7 The least squares fit using Python.

4.3 Linear Rectification

Linear rectification is an approach for converting a nonlinear equation to a linear form. This method helps estimate parameters and constants of equations by using the properties of a straight line.

Example 4.3 Given here is a formula to model y. In this formula, a and b are unknown values. Apply linear rectification to estimate a and b.

$$y = a + x^b$$

Answer:

$$\ln y = \ln a + \ln x^b$$

$$\ln y = \ln a + b \ln x$$

$$\hat{y} = \ln y$$

$$x = \ln x$$

$$\hat{\beta}_1 = b$$

$$\hat{\beta}_0 = \ln a$$

The probability plot is a graphical way to determine whether data follows a certain theoretical distribution. If the data fit well to the assumed distribution, the data points should be close to the straight line in a plot. If there are significant departures from a straight line, it indicates the assumed distribution is not a good fit for the data. The probability plot also enables us to estimate the parameters of the assumed distribution by considering the intercept and slope of the straight line.

4.4 Exponential Distribution Plotting

The linear rectification and probability plot approaches could be applied to the exponential distribution. Based on the equation of the cumulative distribution function of the exponential distribution, we can conduct linear rectification.

$$F(t) = 1 - e^{-\frac{t}{\text{MTTF}}}$$

$$e^{-\frac{t}{\text{MTTF}}} = 1 - F(t)$$

$$-\frac{t}{\text{MTTF}} = \ln\left[1 - F(t)\right]$$

$$t = -\text{MTTF} \cdot \ln\left[1 - F(t)\right] \tag{4.7}$$

After the linear rectification, the parameters can be estimated. The slope of the straight line is an estimation of MTTF.

$$\hat{y} = t$$

$$x = -\ln\left[1 - F(t)\right]$$

$$\hat{\beta}_1 = \text{MTTF}$$

$$\hat{\beta}_0 = 0 \tag{4.8}$$

An alternative approach can be used to estimate the λ:

$$F(t) = 1 - e^{-\frac{t}{\text{MTTF}}}$$

$$-\left[1 - F(t)\right] = -e^{-\frac{t}{\text{MTTF}}}$$

$$-\ln\left[1 - F(t)\right] = \frac{t}{\text{MTTF}} = \lambda t \tag{4.9}$$

After the linear rectification, the parameters can be estimated. The slope of the straight line is an estimation of λ.

$$\hat{y} = -\ln\left[1 - F(t)\right] = \ln\frac{1}{1 - F(t)}$$

$$x = t$$

$$\hat{\beta}_1 = \lambda$$

$$\hat{\beta}_0 = 0 \tag{4.10}$$

Median rank estimates can be used to estimate the cumulative distribution function (CDF) of each failure time. The failure data are initially sorted in order of failure as seen in Table 4.2. There are 10 components, and each component has an equal chance of failure (1/10).

Table 4.2 Cumulative distribution function estimates of failure times.

Time to Failure (hours)	Order of Failure	*f*(*t*)	*F*(*t*)
3	1	1/10	1/10
6	2	1/10	2/10
11	3	1/10	3/10
12	4	1/10	4/10
15	5	1/10	5/10
17	6	1/10	6/10
19	7	1/10	7/10
20	8	1/10	8/10
23	9	1/10	9/10
24	10	1/10	10/10

There is a limitation of this approach to estimating the CDF. As seen in Figure 4.8, the CDF estimate of the last failure was 1 or 100%. It may not be realistic to assume that 100% of components would fail based on the limited sample data.

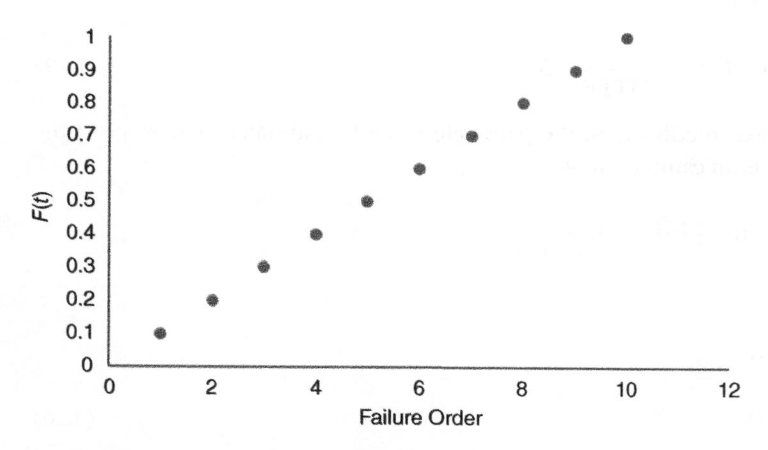

Figure 4.8 The CDF estimates by failure order.

An alternative approach is to use a median rank to estimate the CDF.

$$0.5 = \sum_{k=j}^{N} \binom{N}{k} Z^k (1 - Z)^{N-k} \tag{4.11}$$

Table 4.3 shows the median rank estimate of each failure. For the last failure, it can be calculated as:

$$0.5 = \sum_{10}^{10} \binom{10}{10} Z^{10} (1-Z)^{10-10} = Z^{10}$$

$$Z = 0.5^{\frac{1}{10}} = 0.93 \tag{4.12}$$

Table 4.3 Median rank estimates of failure times.

Time to Failure (hours)	Order of Failure	*f(t)*	Median Rank *F(t)*
3	1	1/10	0.067
6	2	1/10	0.162
11	3	1/10	0.259
12	4	1/10	0.355
15	5	1/10	0.452
17	6	1/10	0.548
19	7	1/10	0.645
20	8	1/10	0.741
23	9	1/10	0.838
24	10	1/10	0.933

Another, simpler way to calculate the median rank estimate is to use Benard's approximation.

$$\hat{F}(t) \cong \frac{r - 0.3}{n + 0.4} \tag{4.13}$$

The r denotes the failure order, and n denotes the total sample size. Table 4.4 summarizes the median rank estimates using Benard's approximation. The results are very similar to the median rank estimates using Equation (4.11).

Table 4.4 Median rank estimates of failure times using Benard's approximation.

Time to Failure (hours)	Order of Failure	*f(t)*	Median Rank *F(t)*
3	1	1/10	0.067
6	2	1/10	0.163

(Continued)

Table 4.4 (Continued)

Time to Failure (hours)	Order of Failure	f(t)	Median Rank F(t)
11	3	1/10	0.260
12	4	1/10	0.356
15	5	1/10	0.452
17	6	1/10	0.548
19	7	1/10	0.644
20	8	1/10	0.740
23	9	1/10	0.837
24	10	1/10	0.933

For example, the fifth failure can be calculated as:

$$\widehat{F}(t) \cong \frac{5-0.3}{10+0.4} = 0.452 \tag{4.14}$$

Example 4.4 Failure testing was conducted for 10 items. At the end of the testing, 6 items failed and 4 items survived. Table 4.5 summarizes the failure of 6 items. The reliability engineer empirically knows that failure data is exponentially distributed. Estimate parameters using a probability plot.

Table 4.5 The time to failure of 6 items.

Time to Failure (hours)	Order of Failure
30	1
60	2
80	3
93	4
120	5
150	6

Answer:
The median rank estimates can be calculated using Benard's approximation. For the first failure:

$$\widehat{F}(t) \cong \frac{1-0.3}{10+0.4} = 0.067$$

Time to Failure (hours)	Order of Failure	Median Rank $F(t)$	$\ln \frac{1}{1-F(t)}$
30	1	0.067	0.070
60	2	0.163	0.178
80	3	0.260	0.301
93	4	0.356	0.440
120	5	0.452	0.601
150	6	0.548	0.794

Here is the linear rectification of the exponential distribution.

$$\ln \frac{1}{1-F(t)} = \lambda t$$

After arranging the x and y data in the probability plot, we can construct a regression line (Figure 4.9).

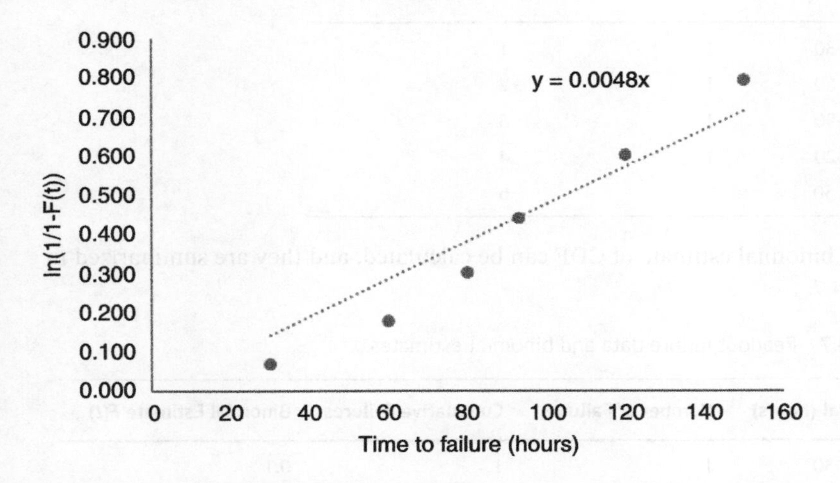

Figure 4.9 A regression line based on the exponential distribution plotting.

The slope of the regression line is 0.0048.

$$\hat{\lambda} = 0.0048$$

The readout (grouped) data is another type of failure data. The number of failures is counted during the interval. This data collection is less expensive compared to obtaining exact failure data.

The binomial estimate is used to estimate the CDF.

$$\hat{F}(T_k) = \frac{\sum_{i=1}^{k} d_i}{n} = \frac{r_k}{n} \tag{4.15}$$

- T_k is the order of readout time.
- d_i is the number of failures during the i_{th} interval.
- n is the sample size.
- r_k is the cumulative number of failures.

Ten components were tested for up to 150 hours to assess their failure characteristics. A readout schedule was considered for every 30 hours as an interval. Table 4.6 summarizes the testing results.

Table 4.6 Readout failure data.

Interval (hours)	Number of Failures	Cumulative Failures
0–30	1	1
30–60	1	2
60–90	1	3
90–120	1	4
120–150	2	6

The binomial estimate of CDF can be calculated, and they are summarized in Table 4.7.

Table 4.7 Readout failure data and binomial estimates.

Interval (hours)	Number of Failures	Cumulative Failures	Binomial Estimate $F(t)$
0–30	1	1	0.1
30–60	1	2	0.2
60–90	1	3	0.3
90–120	1	4	0.4
120–150	2	6	0.6

For the first interval's failure:

$$\hat{F}(T_1) = \frac{1}{10} = 0.1 \tag{4.16}$$

To construct a probability plot of the exponential distribution, the transformed CDF can be calculated as seen in Table 4.8. The readout interval can be divided into start and end times.

Table 4.8 Readout failure data and transformed CDF.

Start Time	End Time	Number of Failures	Cumulative Failures	Binomial Estimate *F*(*t*)	Ln [1/1−*F*(*t*)]
0	30	1	1	0.1	0.105
30	60	1	2	0.2	0.223
60	90	1	3	0.3	0.357
90	120	1	4	0.4	0.511
120	150	2	6	0.6	0.916

The probability plot is finally constructed as seen in Figure 4.10. The transformed CDF is *x* data, and end time is *y* data. The slope value estimates the MTTF as 191.98 hours.

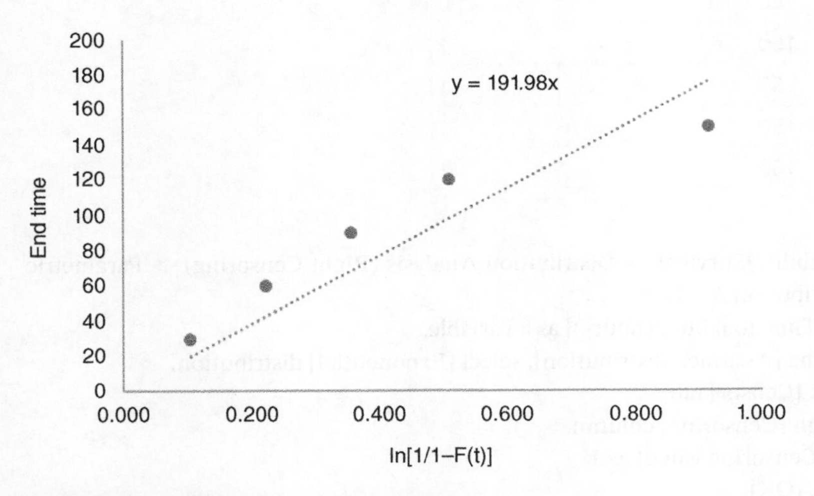

Figure 4.10 Probability plot using readout data.

4.4.1 Excel Practice

The median rank estimates can be calculated using the Excel function.

- BETAINV(0.5, r, n-r+1)

The inverse beta function is used. The *r* indicates the order of failure, and *n* denotes the total sample size.

4.4.2 Minitab Practice

The data of Example 4.4 can be analyzed using Minitab. As seen in Figure 4.11, a censoring column can be added. The 0 indicates no censoring, and 1 denotes the censored values. Since there were 4 survivals after the testing, censoring values of 1 were assigned to them.

↓	C1	C2
	Time to failure (hours)	Censoring
1	30	0
2	60	0
3	80	0
4	93	0
5	120	0
6	150	0
7	150	1
8	150	1
9	150	1
10	150	1

Figure 4.11 Minitab worksheet.

- Reliability/Survival -> Distribution Analysis (Right Censoring) -> Parametric Distribution Analysis
- Set [Time to failure (hours)] as a variable.
- For the [Assumed distribution], select [Exponential] distribution.
- Click [Censor] tab.
- Assign [Censoring] column.
- Set [Censoring value] as 1.
- Click [OK].
- Select [Estimate] tab.

- Choose [Least Squares] as an [Estimation Method].
- Click [OK].

The probability plot will be constructed as seen in Figure 4.12. Since there were some departures of the data points from a straight line, the exponential distribution may not be the best fit in this case.

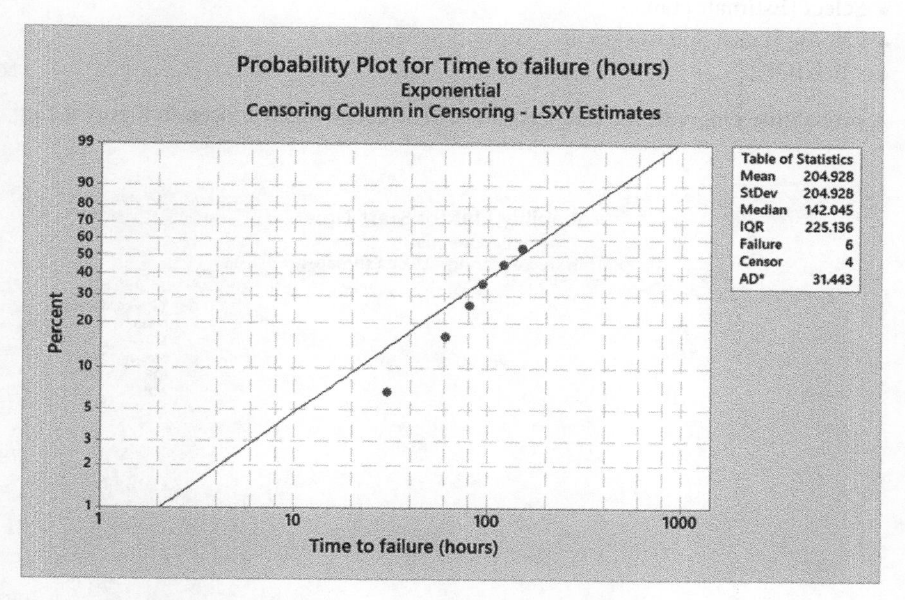

Figure 4.12 Probability plot of the exponential distribution using Minitab.

The readout data of Table 4.4 can also be assessed using Minitab. The data can be arranged as seen in Figure 4.13. Since there were 4 survivals after 150 hours, it could be written as 4 failures in the [150 to *] interval.

↓	C1	C2	C3
	Start Time	End Time	Number of Failures
1	0	30	1
2	30	60	1
3	60	90	1
4	90	120	1
5	120	150	2
6	150	*	4

Figure 4.13 Minitab data worksheet.

- Reliability/Survival -> Distribution Analysis (Arbitrary Censoring) -> Parametric Distribution Analysis
- Set [Start variables] as [Start Time].
- Set [End variables] as [End Time].
- Set [Frequency columns] as [Number of Failures].
- For the [Assumed distribution], select [Exponential] distribution.
- Select [Estimate] tab.
- Choose [Least Squares] as an [Estimation Method].
- Click [OK].

A probability plot with the readout data will be constructed as seen in Figure 4.14.

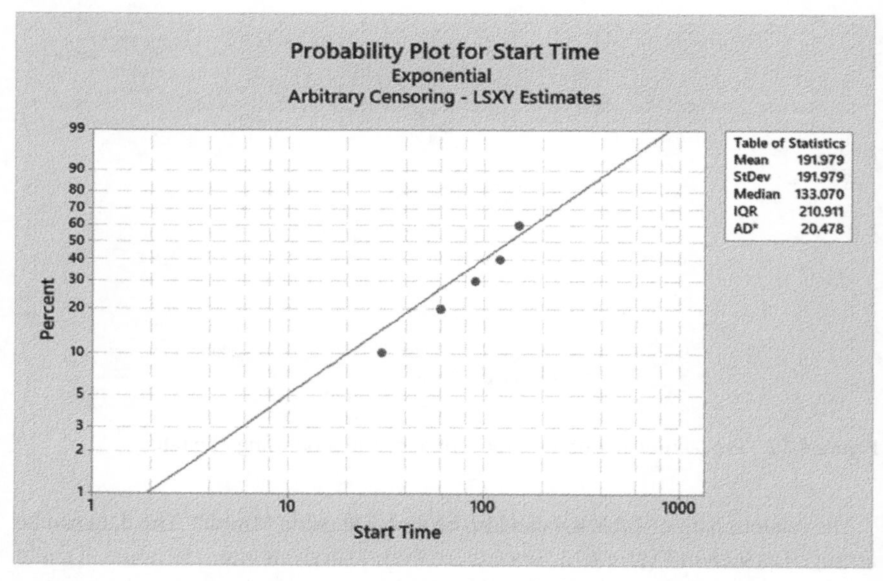

Figure 4.14 Probability plot of the exponential distribution using Minitab.

4.4.3 Python Practice

The Python codes used for constructing the exponential distribution probability plot with exact failure times and readout data are given in Figure 4.15.

The **[surpyval]** library (Knife, 2021) can be installed to perform the analysis.

[x] shows the time to failure data, and **[c]** indicates the censoring of each value. **[1]** means that the data is censored.

[surv.Exponential.fit(x,c)] will run the exponential probability plot based on the data.

[np.histogram(x, bins)] is used to determine the readout schedule of the test.

[np.vstack] is to stack the data as a single array.

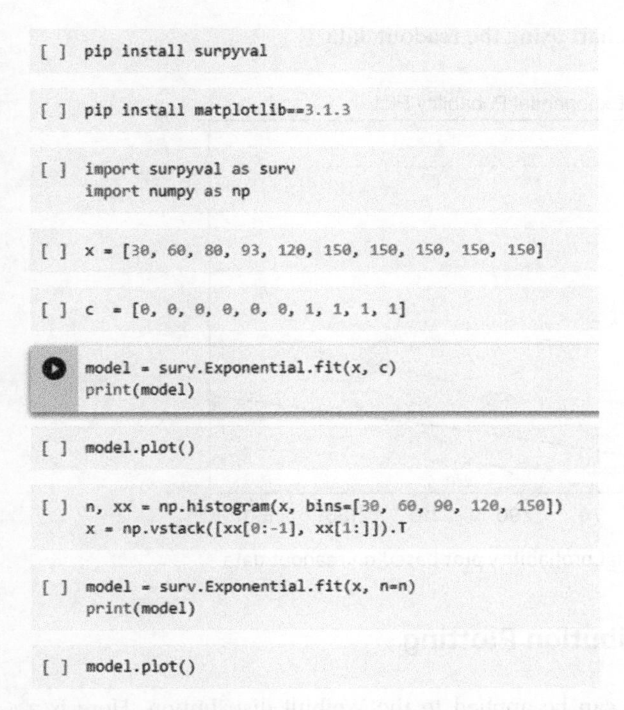

```
[ ] pip install surpyval

[ ] pip install matplotlib==3.1.3

[ ] import surpyval as surv
    import numpy as np

[ ] x = [30, 60, 80, 93, 120, 150, 150, 150, 150, 150]

[ ] c = [0, 0, 0, 0, 0, 0, 1, 1, 1, 1]

    model = surv.Exponential.fit(x, c)
    print(model)

[ ] model.plot()

[ ] n, xx = np.histogram(x, bins=[30, 60, 90, 120, 150])
    x = np.vstack([xx[0:-1], xx[1:]]).T

[ ] model = surv.Exponential.fit(x, n=n)
    print(model)

[ ] model.plot()
```

Figure 4.15 Python codes used for constructing the exponential distribution probability plot.

After running all the codes, the result using the exact failure time data is shown in Figure 4.16.

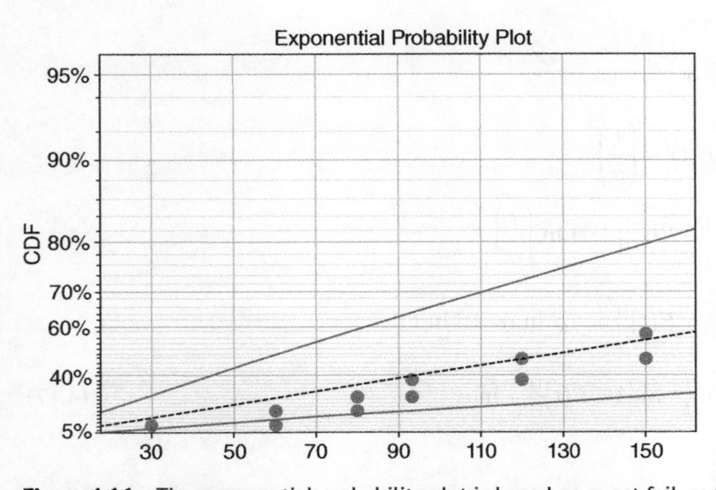

Figure 4.16 The exponential probability plot is based on exact failure times.

Figure 4.17 shows the chart using the readout data.

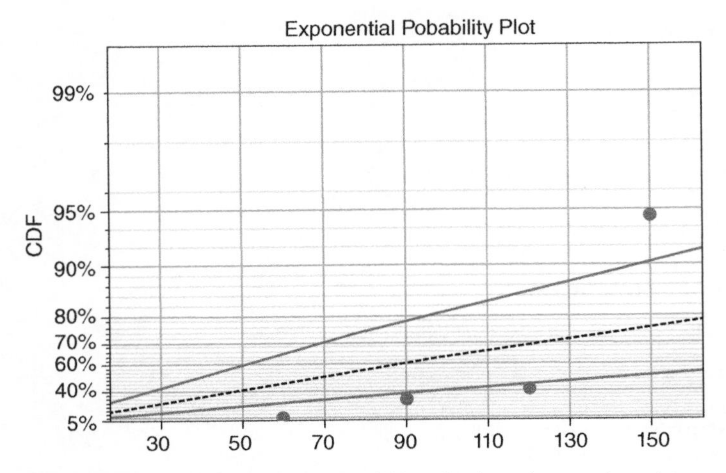

Figure 4.17 The exponential probability plot based on readout data.

4.5 Weibull Distribution Plotting

The probability plotting can be applied to the Weibull distribution. Here is a procedure for conducting a linear rectification based on the CDF formula of the Weibull distribution.

$$F(t) = 1 - e^{-\left(\frac{t}{\alpha}\right)^{\beta}}$$

$$1 - F(t) = e^{-\left(\frac{t}{\alpha}\right)^{\beta}}$$

$$-\ln\left[1 - F(t)\right] = \left(\frac{t}{\alpha}\right)^{\beta}$$

$$\ln\left\{-\ln\left[1 - F(t)\right]\right\} = \beta \ln\left(\frac{t}{\alpha}\right)$$

$$\ln\left\{-\ln\left[1 - F(t)\right]\right\} = -\beta \ln \alpha + \beta \ln t$$

$$\ln \mathrm{t} = \frac{1}{\beta} \ln\left\{-\ln\left[1 - F(t)\right]\right\} + \ln \alpha \tag{4.17}$$

After the linear rectification, the parameters can be estimated.

$$\hat{y} = \ln t$$
$$x = \ln\left\{-\ln\left[1 - F(t)\right]\right\}$$
$$\hat{\beta}_1 = \frac{1}{\beta}$$
$$\hat{\beta}_0 = \ln \alpha \tag{4.18}$$

Example 4.5 Twenty components were tested for up to 200 hours to understand their failure characteristics. At the end of the testing, 17 components failed, and 3 components survived (Table 4.9). The reliability engineer wants to determine whether the Weibull distribution is a reasonable fit for the data.

Table 4.9 Time to failure data of 17 components.

Time to Failure (hours)	Failure Order
2	1
5	2
10	3
13	4
16	5
19	6
20	7
22	8
25	9
60	10
65	11
90	12
130	13
150	14
170	15
180	16
200	17

Answer:

The median rank can be estimated using Benard's approximation. The transformed CDF can be calculated using the median rank estimates.

Time to Failure (hours)	Failure Order	Median Rank $F(t)$	Ln{–ln [1 – $F(t)$]}
2	1	0.034	–3.355
5	2	0.083	–2.442
10	3	0.132	–1.952
13	4	0.181	–1.609
16	5	0.230	–1.340
19	6	0.279	–1.116
20	7	0.328	–0.921
22	8	0.377	–0.747
25	9	0.426	–0.587
60	10	0.475	–0.438
65	11	0.525	–0.297
90	12	0.574	–0.160
130	13	0.623	–0.026
150	14	0.672	0.107
170	15	0.721	0.243
180	16	0.770	0.384
200	17	0.819	0.535

The transformed CDF can be the x data, and $\ln(t)$ can be the y data. The probability plot of the Weibull distribution can be constructed as seen in Figure 4.18.

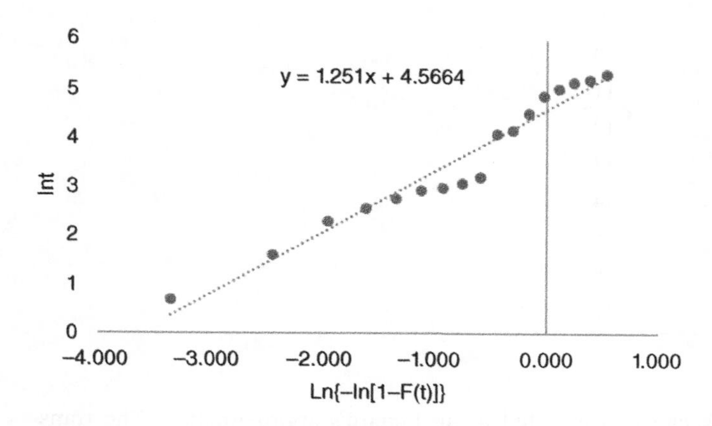

Figure 4.18 The Weibull probability plot.

The characteristics life can be calculated:

$$\hat{\beta}_0 = \ln \alpha = 4.5664$$

$$\alpha = e^{4.5664} = 96.20 \text{ hours}$$

The shape parameter can be calculated:

$$\hat{\beta}_1 = \frac{1}{\beta} = 1.251$$

$$\beta = \frac{1}{1.251} = 0.799$$

4.5.1 Minitab Practice

For the Minitab application, the data can be summarized as seen in Figure 4.19.

↓	C1 Time to failure (hours)	C2 Censoring
1	2	0
2	5	0
3	10	0
4	13	0
5	16	0
6	19	0
7	20	0
8	22	0
9	25	0
10	60	0
11	65	0
12	90	0
13	130	0
14	150	0
15	170	0
16	180	0
17	200	0
18	200	1
19	200	1
20	200	1

Figure 4.19 The data set using Minitab.

The probability plot of the Weibull distribution can be constructed (Figure 4.20). In the table of statistics, the shape indicates the shape parameter of 0.799. The scale denotes the characteristic life of 96.2 hours. These values are very similar to the values calculated from Excel.

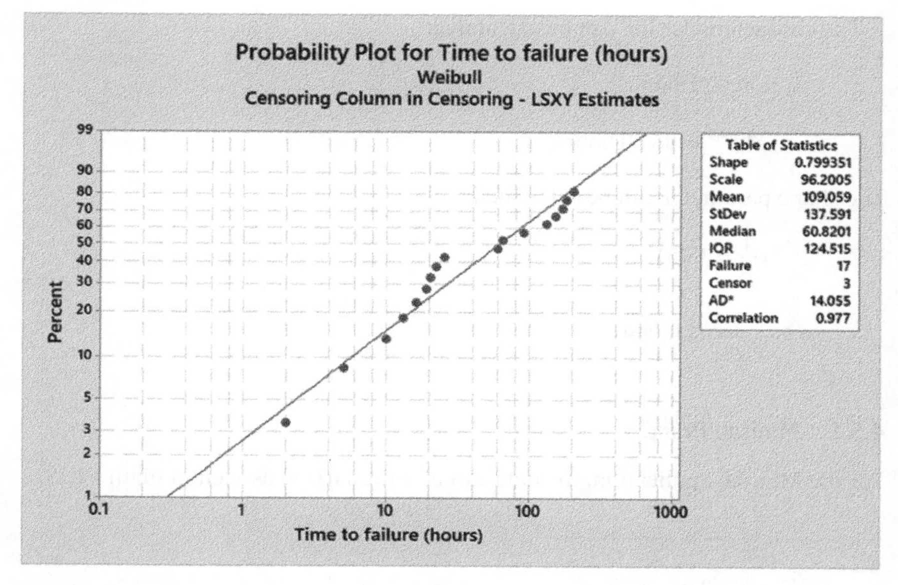

Figure 4.20 The Weibull probability plot using Minitab.

4.5.2 Python Practice

Figure 4.21 shows the Python codes to construct the Weibull distribution probability plot.

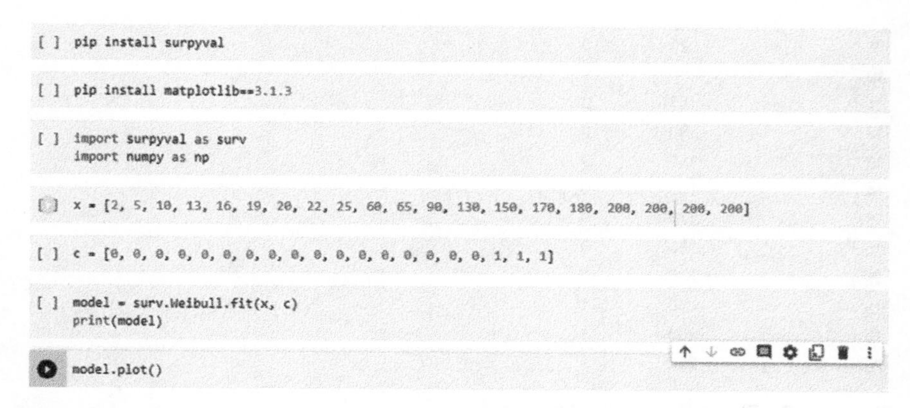

Figure 4.21 The Python codes used to construct the Weibull probability plot.

After running all the codes, the chart is created as seen in Figure 4.22.

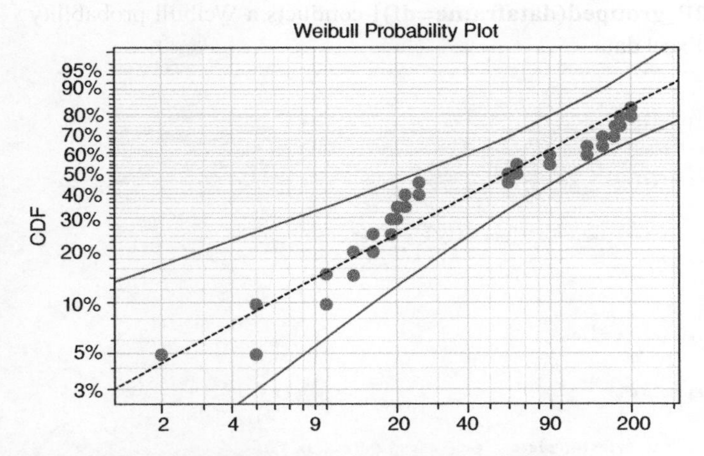

Figure 4.22 The Weibull probability plot using Python.

An alternative way is to directly import the Excel file in a local drive (Figure 4.23).

	A	B	C
1	time	quantity	category
2	2	1	F
3	5	1	F
4	10	1	F
5	13	1	F
6	16	1	F
7	19	1	F
8	20	1	F
9	22	1	F
10	25	1	F
11	60	1	F
12	65	1	F
13	90	1	F
14	130	1	F
15	150	1	F
16	170	1	F
17	180	1	F
18	200	1	F
19	200	3	C

Figure 4.23 The Excel data set used for Python analysis.

[Fit_Weibull_2P_grouped] accepts a data frame for efficiently handling the data from the file (Figure 4.24).

[pandas] is a library for data manipulation and analysis.

[io] allows managing the data file.

[files.upload()] helps to upload the file in the computer.

[pd.read_excel] is used to read the Excel file data.

[Fit_Weibull_2P_grouped(dataframe=df)] conducts a Weibull probability plot based on the Excel data.

```
pip install reliability
```

```
[ ] pip install matplotlib==3.1.3
```

```
[ ] from reliability.Fitters import Fit_Weibull_2P_grouped
    import pandas as pd
```

```
[ ] import io
    from google.colab import files
```

```
[ ] uploaded = files.upload()
```

```
[ ] df = pd.read_excel(io.BytesIO(uploaded.get('WeibullData.xlsx')))
```

```
[ ] print(df.head(21),'\n')
    Fit_Weibull_2P_grouped(dataframe=df)
```

Figure 4.24 The Python codes used to create the Weibull probability plot.

After running all the codes, the Weibull probability plot is created as shown in Figure 4.25.

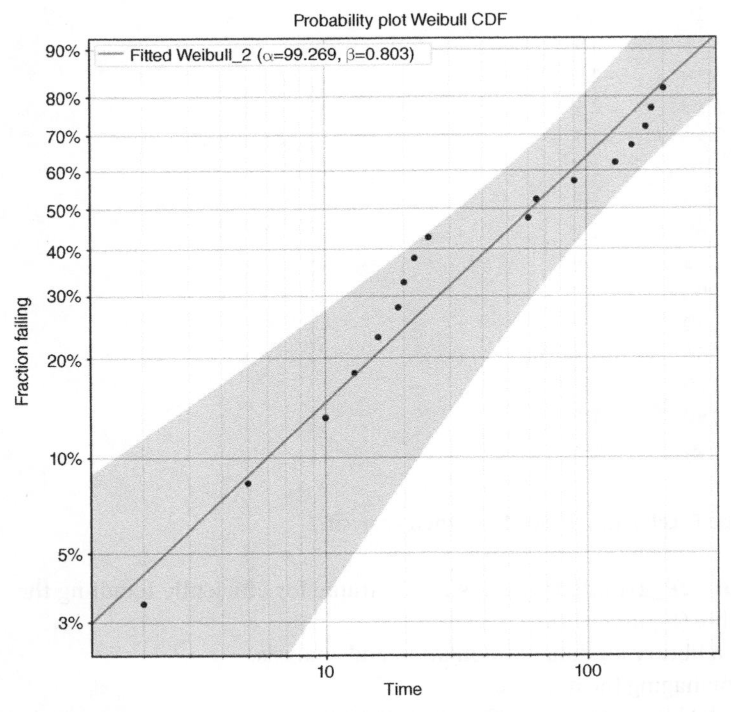

Figure 4.25 The Weibull probability plot with Python.

4.6 Normal Distribution Plotting

The probability plot approach can be applied to the normal distribution. The linear rectification is conducted based on the standard normal distribution.

$$F(t) = \Phi(z)$$

$$z = \Phi^{-1}(F(t))$$

$$\frac{t-\mu}{\sigma} = \Phi^{-1}(F(t))$$

$$t = \sigma \cdot \Phi^{-1}(F(t)) + \mu \tag{4.19}$$

After the linear rectification, the parameters can be estimated.

$$\hat{y} = t$$

$$x = \Phi^{-1}(F(t)) \text{ or } Z$$

$$\hat{\beta}_1 = \sigma$$

$$\hat{\beta}_0 = \mu \tag{4.20}$$

Example 4.6 Ten components were under a life test, and they were observed until they failed. Table 4.10 shows the exact failure time of each component. The reliability engineer assumes a normal distribution of the failure characteristics. Construct a probability plot and estimate parameters.

Table 4.10 Time to failure data of 10 components.

Time to Failure (days)	Order of Failure
35	1
45	2
55	3
58	4
68	5
68	6
72	7
73	8
75	9
80	10

Answer:
The median rank estimates can be calculated using Benard's approximation. The Z values can be calculated using the Excel function: NORMSINV(Median rank estimates).

Time to Failure (days)	Order of Failure	Median Rank	Z
35	1	0.067	−1.496
45	2	0.163	−0.980
55	3	0.260	−0.645
58	4	0.356	−0.370
68	5	0.452	−0.121
68	6	0.548	0.121
72	7	0.644	0.370
73	8	0.740	0.645
75	9	0.837	0.980
80	10	0.933	1.496

The probability plot was constructed using Excel (Figure 4.26). The Z values were assigned to x data, and time to failure was assigned to the y data. Here are the estimations of the parameters.

- $\hat{\beta}_1 = \sigma = 15.07\,\text{days}$
- $\hat{\beta}_0 = \mu = 62.9\,\text{days}$

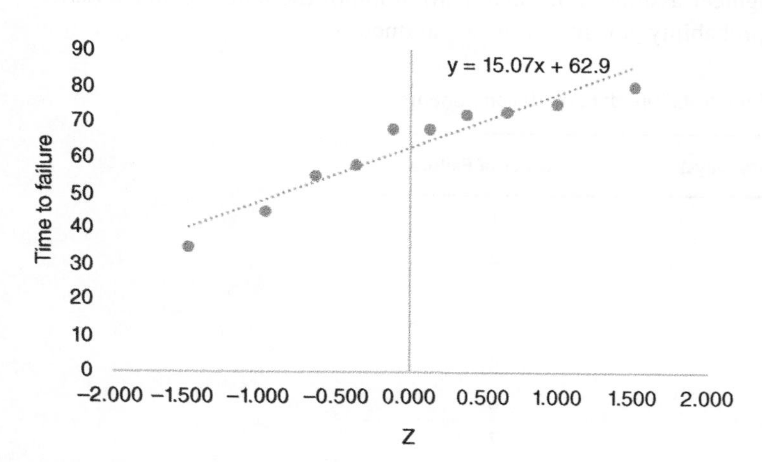

Figure 4.26 The normal probability plot.

4.6.1 Minitab Practice

Minitab can also be used to construct a probability plot (Figure 4.27). In the table of statistics, the mean was 62.9 days, and StDev was 15.07 days. These figures are very similar to the parameter estimates using Excel.

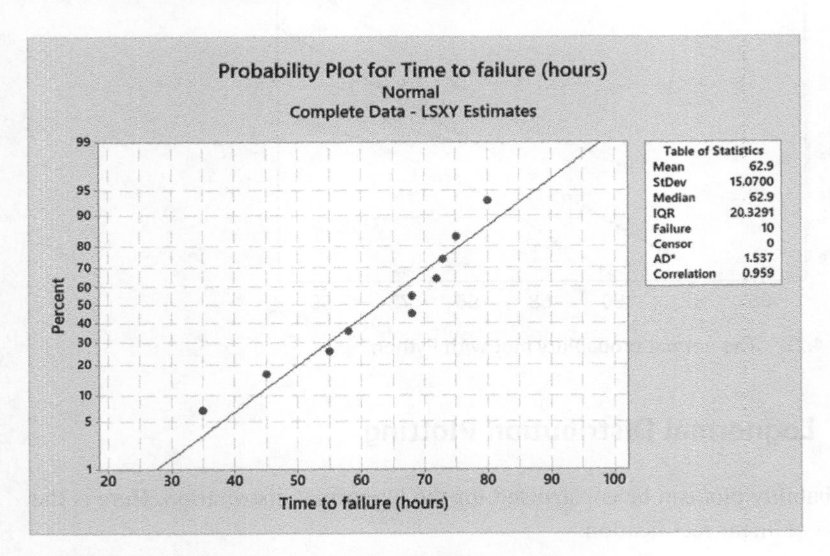

Figure 4.27 The normal probability plot using Minitab.

4.6.2 Python Practice

The Python codes to construct the normal distribution probability plot are shown in Figure 4.28.

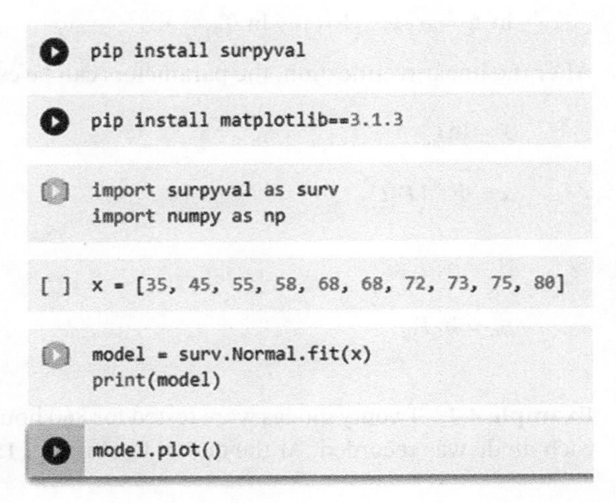

pip install surpyval
pip install matplotlib==3.1.3
import surpyval as surv
import numpy as np
[ ] x = [35, 45, 55, 58, 68, 68, 72, 73, 75, 80]
model = surv.Normal.fit(x)
print(model)
model.plot()

Figure 4.28 The Python codes used to construct the normal probability plot.

After running all the codes, we construct the chart (Figure 4.29).

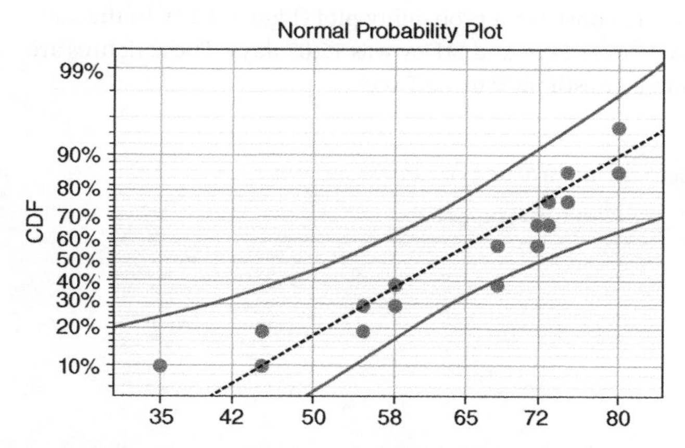

Figure 4.29 The normal probability plot with Python.

4.7 Lognormal Distribution Plotting

A probability plot can be constructed for the lognormal distribution. Here is the process of linear rectification.

$$F(t) = \Phi(z)$$

$$z = \Phi^{-1}(F(t))$$

$$\frac{\ln t_a - \ln T_{50}}{\sigma} = \Phi^{-1}(F(t))$$

$$\ln t_a = \sigma \cdot \Phi^{-1}(F(t)) + \ln T_{50} \tag{4.21}$$

After the linear rectification, the parameters can be estimated.

$$\hat{y} = \ln t_a$$

$$x = \Phi^{-1}(F(t))$$

$$\hat{\beta}_1 = \sigma$$

$$\hat{\beta}_0 = \ln T_{50} \tag{4.22}$$

Example 4.7 Twenty diodes were tested for 860 hours. The exact failure time of each diode was recorded. At the end of the testing, 15 diodes failed (Table 4.11).

The lognormal distribution is assumed for the failure data. Construct a probability plot and estimate the parameters of the lognormal distribution.

Table 4.11 Time to failure data of 15 components.

Time to Failure (hours)	Order of Failure
4	1
24	2
60	3
80	4
150	5
160	6
220	7
230	8
280	9
290	10
480	11
690	12
720	13
775	14
860	15

Answer:
The median rank estimates are calculated. The *Z* values are calculated using the Excel function NORMSINV.

Time to Failure (hours)	Order of Failure	Median Rank Estimates	*Z*
4	1	0.034	−1.821
24	2	0.083	−1.383
60	3	0.132	−1.115
80	4	0.181	−0.910
150	5	0.230	−0.738
160	6	0.279	−0.585
220	7	0.328	−0.444
230	8	0.377	−0.312
280	9	0.426	−0.185
290	10	0.475	−0.061

Time to Failure (hours)	Order of Failure	Median Rank Estimates	Z
480	11	0.525	0.061
690	12	0.574	0.185
720	13	0.623	0.312
775	14	0.672	0.444
860	15	0.721	0.585

The probability plot was constructed using Excel (Figure 4.30). The Z values were assigned to x data, and $\ln t$ values were set as y data. Here are the estimates of the parameters.

- $\hat{\beta}_1 = \sigma = 2.04$ hours
- $\hat{\beta}_0 = \ln T_{50} = 6.0105$
- $T_{50} = e^{6.0105} = 407.69$ hours

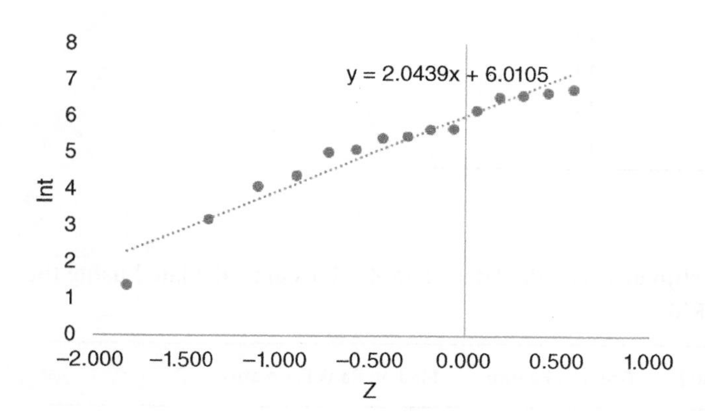

Figure 4.30 The lognormal probability plot.

4.7.1 Minitab Practice

Minitab can be used to construct a probability plot. The data can be set as seen in Figure 4.31.

Figure 4.31 The data set using Minitab.

↓	C1	C2	C3
	Time to failure (hours)	The order of failure	Censoring
1	4	1	0
2	24	2	0
3	60	3	0
4	80	4	0
5	150	5	0
6	160	6	0
7	220	7	0
8	230	8	0
9	280	9	0
10	290	10	0
11	480	11	0
12	690	12	0
13	720	13	0
14	775	14	0
15	860	15	0
16	860	16	1
17	860	17	1
18	860	18	1
19	860	19	1
20	860	20	1

The probability plot using Minitab is shown in Figure 4.32. In the table of statistics, Scale indicates σ of 2.04, and the median denotes T_{50} of 407.7 hours. The estimates are similar to the estimates found using Excel.

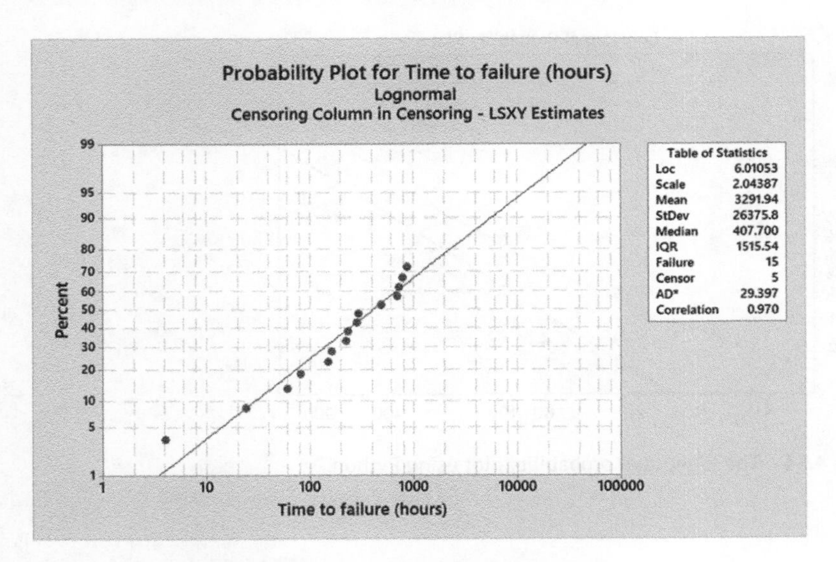

Figure 4.32 The lognormal probability plot using Minitab.

4.7.2 Python Practice

The Python codes used to construct the lognormal distribution probability plot are shown in Figure 4.33).

```
[ ] pip install surpyval
```

```
[ ] pip install matplotlib==3.1.3
```

```
[ ] import surpyval as surv
    import numpy as np
```

```
[ ] x = [4, 24, 60, 80, 150, 160, 220, 230, 280, 290, 480, 690, 720, 775, 860]
```

```
[ ] model = surv.LogNormal.fit(x)
    print(model)
```

```
model.plot()
```

Figure 4.33 The Python codes to construct the lognormal probability plot.

After running all codes, the chart is constructed as shown in Figure 4.34.

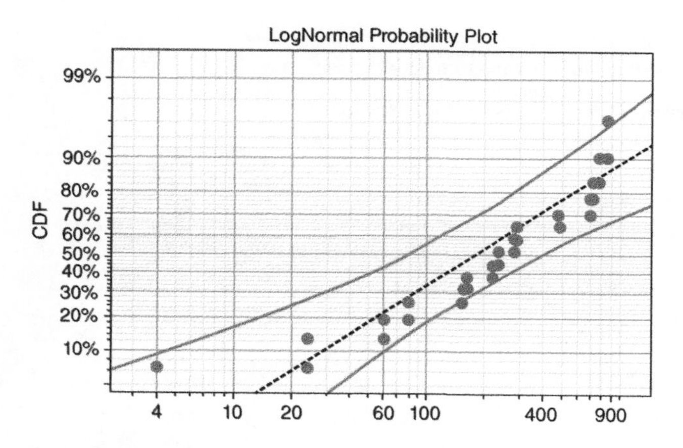

Figure 4.34 The lognormal probability plot using Python.

4.8 Summary

- Graphical procedures could be useful to validate the assumptions of particular distributions to the data.
- Reliability data plotting also helps to estimate parameters of the distributions quickly.
- The straight line equation explains that the changes in y are proportional to the changes in x. The slope parameter, β_1, can also be considered a proportionality factor.
- The least squares fit determines the fitted line showing the minimum of the sum of the squares of the deviations from the line.
- The linear rectification and probability plot approaches could be applied to various life distributions including exponential, Weibull, normal, and lognormal distributions.

Exercises

1 Conduct a linear rectification of this model: $y = ax^3 + b$. Assign the parameters into a linear form.

2 Ten components were tested to assess their failure characteristics. At the end of testing, all components had failed. The Weibull distribution was assumed for this failure data. Construct a probability plot and estimate the parameters of the Weibull distribution.

Time to Failure (days)	Failure Order
3	1
6	2
11	3
14	4
17	5
20	6
21	7
23	8
26	9
61	10

3 Ten components were tested to assess their failure characteristics. Readout data were collected for every 10 days' intervals. The Weibull distribution was assumed for this failure data. Construct a probability plot and estimate the parameters of the Weibull distribution. Compare the parameters with the results of Exercise 2.

Interval (days)	Number of Failures
0–10	3
10–20	18
20–30	24
30–40	0
40–50	0
50–60	0
60–70	1

4 Ten transistors were tested to evaluate their life characteristics. At the end of testing, all transistors failed. The reliability engineer assumes that the failure characteristics could be related to the exponential or Weibull distribution. Conduct a probability plot, and determine which distribution would be the best fit to this data. After determining an adequate distribution, estimate the parameters.

Time to Failure (days)	Failure Order
4	1
7	2
12	3
13	4
16	5
18	6
20	7
21	8
24	9
25	10

5 Construct the regression line formula based on the coordinate data of x and y.

x	y
1.2	3.1
2.1	4.4
3.3	5.5
4.2	6.3
5.3	7.6

Reference

Knife, D. (2021). SurPyval: Survival analysis with Python. *Journal of Open Source Software* 6 (64):3484.

5

Accelerated Life Testing

Chapter Overview and Learning Objectives

- To understand the objectives of conducting accelerated life testing.
- To understand the assumptions of linear acceleration models.
- To apply various lifetime distributions to the linear acceleration models.
- To understand the objectives of the Arrhenius model.
- To apply Minitab to conduct accelerated life testing.

5.1 Accelerated Testing Theory

It is not uncommon to see highly reliable components these days. If we test highly reliable components in normal conditions, we may not obtain sufficient failure data. We may need to stress components in extreme conditions to observe failure. This is a motivation for conducting accelerated testing. Based on the failure data from high-stress conditions, we project failure characteristics of components in normal or use conditions.

For example, a humidified laboratory oven can be utilized to elevate temperature and humidity. In this extreme environment, time to corrosion failure would be substantially shortened. In such a case, time to failure would be accelerated.

Figure 5.1 illustrates accelerated testing theory. The horizontal axis shows different stress levels, and the vertical axis shows the time to failure. The failure characteristics are related to the normal distribution. As seen in Figure 5.1, the mean time to failure (MTTF) at the use condition is much longer than the MTTF in the high-stress condition. Nevertheless, there are no changes in the distribution shape between the two stress conditions. In accelerated testing theory, it is assumed that the same failure mechanism holds through various stress levels. The failure time is only expected to be accelerated.

Reliability Analysis Using MINITAB and Python, First Edition. Jaejin Hwang.
© 2023 John Wiley & Sons, Inc. Published 2023 by John Wiley & Sons, Inc.
Companion Website: www.wiley.com\go\Hwang\ReliabilityAnalysisUsingMinitabandPython

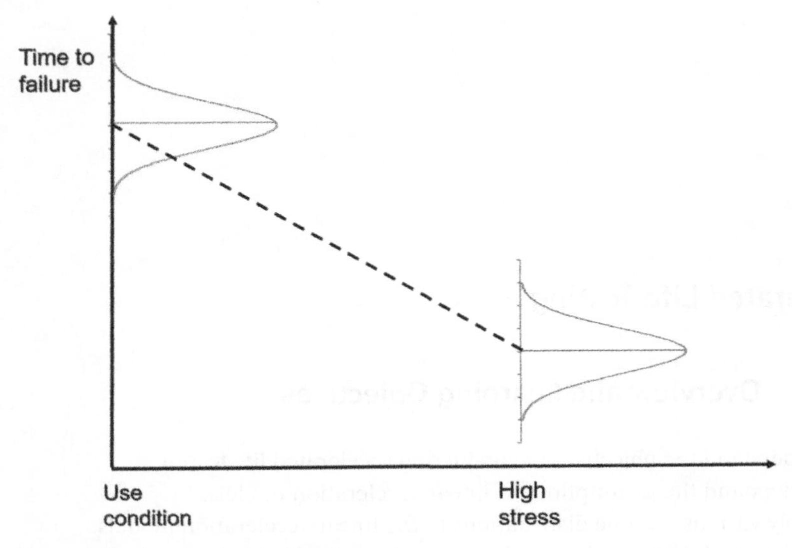

Figure 5.1 Accelerated testing theory.

The linear acceleration model describes the mathematical relationship between stress levels and life characteristics. The equation can be written as

$$t_u = AF \times t_s \tag{5.1}$$

The t_u denotes the time to failure at the use condition. The t_s represents the time to failure at stress conditions. *AF* is an acceleration factor (constant value). This linear model can be applied to the cumulative distribution function.

$$F_u(t_u) = F_s(t_s) = F_s\left(\frac{t_u}{AF}\right) \tag{5.2}$$

Example 5.1 A group of transistors have a mean time to failure (MTTF) of 500 days at the use condition. In the stress condition at 100 °C, the MTTF is 50 days. The failure characteristics are expected to be consistent in a stress condition. Calculate the acceleration factor.

Answer:

$$t_u = AF \times t_s$$

$$AF = \frac{t_u}{t_s} = \frac{500}{50} = 10$$

Table 5.1 shows various failure mechanism and their related stresses used for accelerated testing.

Table 5.1 Various failure mechanisms and accelerated stresses used in testing.

Failure Mechanism	Accelerated Stresses
Corrosion	• Temperature • Humidity
Stress–strength	• Load magnitude • Load frequency
Adhesive wear	• Surface area • Friction • Loading • Duration
Mechanical fatigue	• Load amplitude • Load frequency • Duration
Thermal crack	• Temperature • Temperature cycling

5.2 Exponential Distribution Acceleration

The accelerated testing theory can be applied to the exponential distribution. The linear acceleration model can be applied.

$$\lambda_u = \frac{\lambda_s}{AF} \tag{5.3}$$

The λ_u denotes the constant failure rate at the use condition, and λ_s means the failure rate at the stress condition.

The cumulative distribution function (CDF) with the linear acceleration model can be written as

$$F_u(t_u) = F_s\left(\frac{t_u}{AF}\right) = 1 - e^{-\lambda_s\left(\frac{t_u}{AF}\right)} = 1 - e^{-\left(\frac{\lambda_s}{AF}\right)t_u} = 1 - e^{-\lambda_u t_u} \tag{5.4}$$

Example 5.2 Diodes were tested in a high-stress environment. The failure characteristics were related to the exponential distribution, and the mean time to failure (MTTF) was 300 hours. An acceleration factor of 30 is known between use and high-stress conditions. Calculate the probability of failure by 1000 hours at the use condition.

Answer:

$$\lambda_u = \frac{\lambda_s}{AF} = \frac{\frac{1}{300}}{30} = 0.00011$$

$$F_u(1000) = 1 - e^{-0.00011 \times 1000} = 0.104$$

Example 5.3 Transistors were tested in a high-temperature condition in a laboratory, and the MTTF was 700 hours. An acceleration factor of 20 is known. Calculate the transistors' MTTF at the use condition, and the probability of survival after 1000 hours at the use condition.

Answer:

$$\lambda_u = \frac{\lambda_s}{AF} = \frac{\frac{1}{700}}{20} = 0.00007$$

$$R_u(1000) = e^{-0.00007 \times 1000} = 0.932$$

5.3 Weibull Distribution Acceleration

The accelerated testing theory can also be applied to the Weibull distribution. Considering the linear acceleration model, the equation is as follows:

$$\alpha_u = AF \times \alpha_s$$

$$\beta_u = \beta_s = \beta \tag{5.5}$$

The α_u represents a characteristics life at the use condition, and α_s is a characteristic life at the stress condition. The β is a shape parameter. Since consistent failure characteristics are assumed between the stress and use conditions, the shape parameter, β, is expected to be the same.

The cumulative distribution function of Weibull distribution acceleration can be written as

$$F_u(t_u) = 1 - e^{-\left(\frac{t_u}{\alpha_u}\right)^{\beta_u}} = 1 - e^{-\left(\frac{t_u}{\alpha_s \times AF}\right)^{\beta_s}} \tag{5.6}$$

The F_u is the CDF at the use condition, and t_u is a time to failure at the use condition. The failure rate function can be calculated using the linear acceleration model.

$$h_u(t_u) = \frac{1}{AF} h_s(t_s) = \frac{1}{AF} h_s\left(\frac{t_u}{AF}\right) = \frac{1}{AF}\left(\frac{\beta_s}{\alpha_s}\right)\left(\frac{t_u}{AF \times \alpha_s}\right)^{\beta_s - 1}$$
$$= \frac{1}{AF^{\beta}} \frac{\beta_s}{\alpha_s}\left(\frac{t_u}{\alpha_s}\right)^{\beta-1} = \frac{h_s(t_u)}{AF^{\beta}} \quad (5.7)$$

The h_u is the failure rate at the use condition, and h_s is the failure rate at the stress condition.

Example 5.4 Components were tested at high temperatures to accelerate failure. The failure characteristics were related to the Weibull distribution with $\alpha = 500$ hours and $\beta = 1.0$. The acceleration factor is assumed to be 100. In the use condition, (1) calculate the failure rate over 20,000 hours, and (2) calculate the probability of failure by 30,000 hours.

Answer:

(1) $h_u(20{,}000) = \frac{1}{AF^{\beta}} \frac{\beta_s}{\alpha_s}\left(\frac{t_u}{\alpha_s}\right)^{\beta-1} = \frac{1}{100^{1.0}} \frac{1.0}{500}\left(\frac{20{,}000}{500}\right)^{1-1} = 0.00002$

(2) $F_u(30{,}000) = 1 - e^{-\left(\frac{30{,}000}{500 \times 100}\right)^{1.0}} = 0.451$

5.3.1 Minitab Practice

Nine components were tested under use (25 °C) and high-stress (85 °C) conditions. The time to failure by 200 hours data was collected (Table 5.2). The Weibull distribution is assumed for the failure data. Construct probability plots to determine whether the Weibull distribution is a proper fit. Estimate the failure characteristics on 50 °C.

The data can be initially arranged for the Minitab application (Figure 5.2).

- Go to Stat > Reliability/Survival > Accelerated Life Testing.

Table 5.2 Time to failure data with use and stress conditions.

Use Condition (25 °C)	Stress Condition (85 °C)
50	5
65	7
75	10
90	15
101	20
125	30
135	50
160	56
188	70

↓	C1	C2
	Failure Time	Temperature
1	50	25
2	65	25
3	75	25
4	90	25
5	101	25
6	125	25
7	135	25
8	160	25
9	188	25
10	5	85
11	7	85
12	10	85
13	15	85
14	20	85
15	30	85
16	50	85
17	56	85
18	70	85

Figure 5.2 Minitab failure time data worksheet.

- Select [Responses are uncens/right censored data.
- Set [Failure Time] as [Variables/Start variables].
- Set [Temperature] as [Accelerating var].
- Set [Weibull] as [Assumed distribution].
- Click the [Graphs] tab.
- Write [50] under [Design value to include on plots].
- Under the [Probability plot for each accelerating level based on fitted model], check [Display no confidence interval]. Hit [OK].
- Click [OK].

The probability plot could be constructed as seen in Figure 5.3. The estimation of failure characteristics at 50 °C was constructed as well. However, the Weibull distribution does not look like an adequate fit. An alternative distribution could be explored.

In Figure 5.3, the same shape parameter ($\beta = 1.73961$) was applied to all stress conditions. If we want to estimate the shape parameter individually for each stress condition, we can modify the option in Minitab.

- Click the [Graphs] tab.
- Under the [Diagnostic Plots], check [Probability plot for each accelerating level based on individual fits].

The probability plot with individual fit can be established as seen in Figure 5.4. This option looks to be a much better fit with the Weibull distribution. This means that the failure characteristics may not be consistent across different temperature conditions.

5.3.2 Python Practice

[failure times_at_stress_1] sets the time to failure data with the temperature 25 °C.

[failure_stress_1] denotes the temperature condition of 25 °C.

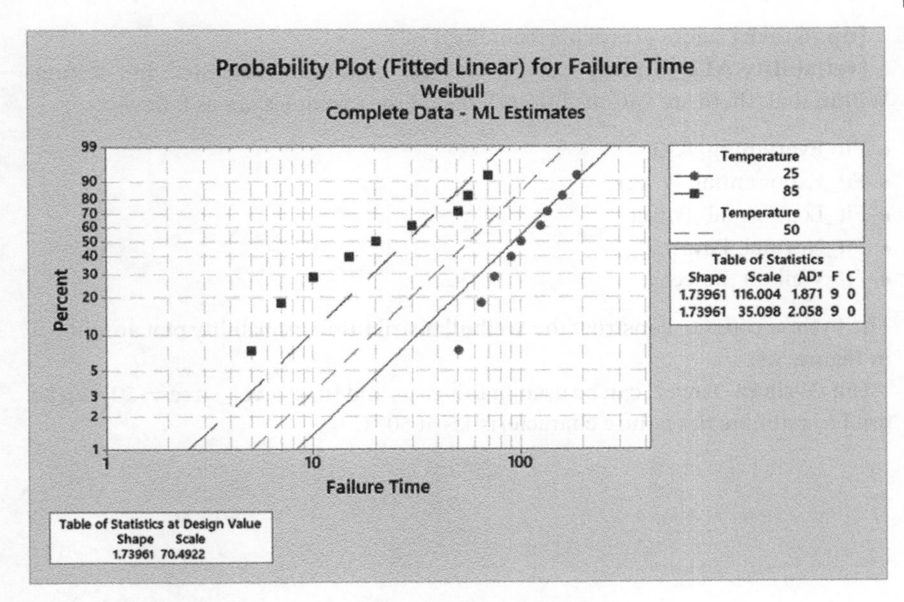

Figure 5.3 Probability plots of accelerated testing using Minitab.

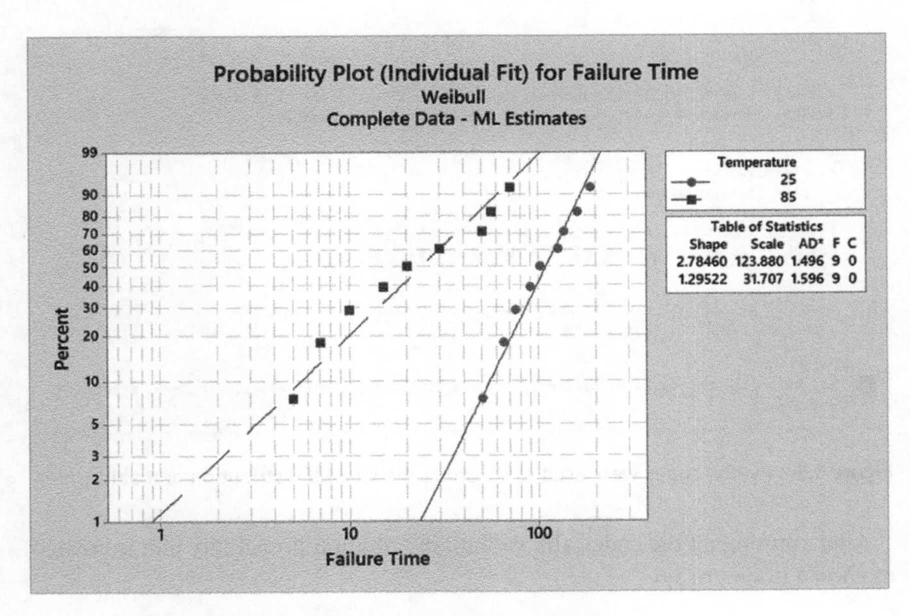

Figure 5.4 Probability plots with individual fit using Minitab.

[np.hstack] stacks arrays horizontally.

[reliability.ALT_fitters] can be used to conduct accelerated life testing. Within that, there are various fitters. Some of the examples are as follows:

- Fit_Everything_ALT
- Fit_Exponential_Power
- Fit_Lognormal_Power
- Fit_Normal_Power
- Fit_Weibull_Power

The Python codes to construct the Weibull distribution probability plot are shown in Figure 5.5.

[Fit_Weibull_Power] can be used in this case, and [use_level_stress= 50] can be used to estimate the failure characteristics at 50 °C.

```
[ ] pip install reliability

[ ] pip install matplotlib==3.1.3

[ ] import numpy as np

[ ] failure_times_at_stress_1 = [50,65,75,90,101,125,135,160,188]
    failure_stress_1 = [25,25,25,25,25,25,25,25,25]
    failure_times_at_stress_2 = [5,7,10,15,20,30,50,56,70]
    failure_stress_2 = [85,85,85,85,85,85,85,85,85]

[ ] failures = np.hstack([failure_times_at_stress_1,failure_times_at_stress_2])
    failure_stresses = np.hstack([failure_stress_1,failure_stress_2])

[ ] print(failures)
    print(failure_stresses)

    [ 50  65  75  90 101 125 135 160 188   5   7  10  15  20  30  50  56  70]
    [25 25 25 25 25 25 25 25 25 85 85 85 85 85 85 85 85 85]

[ ] from reliability.ALT_fitters import Fit_Weibull_Power
    import matplotlib.pyplot as plt

    Fit_Weibull_Power(failures=failures, failure_stress=failure_stresses, use_level_stress=50)
    plt.show()
```

Figure 5.5 Python codes were used to construct the Weibull distribution probability plot.

After running all the codes, the Weibull distribution probability plot is created as shown in Figure 5.6.

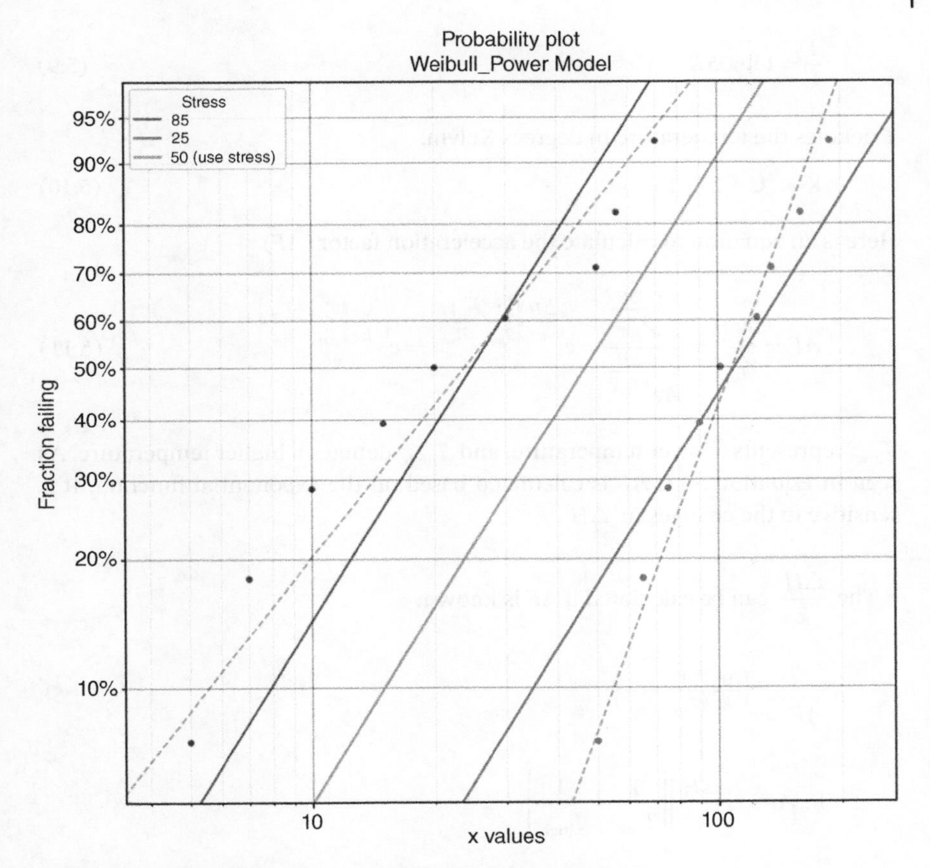

Figure 5.6 The Weibull distribution probability plot with Python.

5.4 Arrhenius Model

The Arrhenius model can be used when failures are accelerated based on an increase in temperature. The Arrhenius model has been commonly used to understand the failure characteristics of electronic components such as batteries, semiconductors, and insulators. The model can be applied to various life distributions.

The model equation is

$$r = Ae^{\frac{\Delta H}{kT}} \tag{5.8}$$

The r represents reaction (e.g., time to failure), and the A is a scaling factor. The $\Delta \mathrm{H}$ is an activation energy for the process (electron volts), and it typically ranges from 0.3 to 1.5. The k is Boltzmann's constant (8.617×10^{-5} eV / K). It can also be written as

$$\frac{1}{k} = 11{,}605\,K\,/\,eV \tag{5.9}$$

T denotes the temperature in degrees Kelvin.

$$\mathrm{K} = {}^{\circ}\mathrm{C} + 273.15 \tag{5.10}$$

Here is an equation to calculate the acceleration factor (AF).

$$AF = \frac{r_{T_{low}}}{r_{T_{high}}} = \frac{Ae^{\frac{\Delta H}{kT_{low}}}}{Ae^{\frac{\Delta H}{kT_{high}}}} = e^{\left(\frac{\Delta H}{k}\right)\left(\frac{1}{T_{low}} - \frac{1}{T_{high}}\right)} = e^{\left(\frac{\Delta H}{k}\right)\left(\frac{T_{high} - T_{low}}{T_{high} \times T_{low}}\right)} \tag{5.11}$$

T_{low} represents a lower temperature, and T_{high} denotes a higher temperature. As seen in Equation 5.11, AF is calculated based on the exponential function. It is sensitive to the changes in ΔH.

The $\frac{\Delta H}{k}$ can be calculated if AF is known.

$$AF = e^{\left(\frac{\Delta H}{k}\right)\left(\frac{1}{T_{low}} - \frac{1}{T_{high}}\right)}$$

$$\ln AF = \left(\frac{\Delta H}{k}\right)\left(\frac{1}{T_{low}} - \frac{1}{T_{high}}\right)$$

$$\frac{\Delta H}{k} = \frac{\ln AF}{\left(\frac{1}{T_{low}} - \frac{1}{T_{high}}\right)} \tag{5.12}$$

Example 5.5 The components' use condition temperature is 25 °C, and the stress condition temperature is 125 °C. The ΔH is known to be 1.0. Calculate the acceleration factor using the Arrhenius model.

Answer:

$$T_{low} = 25 + 273.15 = 298.15$$

$$T_{high} = 125 + 273.15 = 398.15$$

$$AF = e^{\left(\frac{\Delta H}{k}\right)\left(\frac{1}{T_{low}} - \frac{1}{T_{high}}\right)} = e^{1.0 \times 11605 \times \left(\frac{1}{298.15} - \frac{1}{398.15}\right)} = 17607$$

Example 5.6 The components' use condition temperature is 35 °C, and the stress condition temperature is 125 °C. The *AF* is known to be 50. Calculate an estimate of $\frac{\Delta H}{k}$.

Answer:

$$T_{low} = 35 + 273.15 = 308.15$$

$$T_{high} = 125 + 273.15 = 398.15$$

$$\frac{\Delta H}{k} = \frac{\ln AF}{\left(\frac{1}{T_{low}} - \frac{1}{T_{high}}\right)} = \frac{\ln 50}{\left(\frac{1}{308.15} - \frac{1}{398.15}\right)} = 5333$$

5.4.1 Minitab Practice

Forty components were tested under use (25 °C), low-stress (85 °C), and high-stress (135 °C) conditions. The number of failures were counted based on the readout schedule (Table 5.3). The Weibull distribution is assumed for the failure data. Construct a probability plot using the Arrhenius model.

Table 5.3 Readout data of failures with use and two different stress conditions.

Start Time	End Time	Use Condition (25 °C)	Low-Stress Condition (85 °C)	High-Stress Condition (135 °C)
0	25	1	5	7
25	50	1	7	8
50	75	0	8	9
75	100	1	3	4
100	125	0	2	3
125	150	3	4	4
150	175	0	3	3
175	200	1	2	2
200	*	33	6	0

The data can be initially arranged for the Minitab application (Figure 5.7).

↓	C1	C2	C3	C4
	Start Time	End Time	Number of failures	Temperature
1	0	25	1	25
2	25	50	1	25
3	50	75	0	25
4	75	100	1	25
5	100	125	0	25
6	125	150	3	25
7	150	175	0	25
8	175	200	1	25
9	200	*	33	25
10	0	25	5	85
11	25	50	7	85
12	50	75	8	85
13	75	100	3	85
14	100	125	2	85
15	125	150	4	85
16	150	175	3	85
17	175	200	2	85
18	200	*	6	85
19	0	25	7	135
20	25	50	8	135
21	50	75	9	135
22	75	100	4	135
23	100	125	3	135
24	125	150	4	135
25	150	175	3	135
26	175	200	2	135
27	200	*	0	135

Figure 5.7 Minitab readout data worksheet.

- Go to Stat > Reliability/Survival > Accelerated Life Testing.
- Select [Responses are uncens/arbitrarily censored data].
- Set [Start Time] as [Variables/Start variables].
- Set [End Time] as [End variables].
- Set [Number of failures] as [Freq. columns].
- Set [Temperature] as [Accelerating var].
- **Set [Relationship] as [Arrhenius].**
- Set [Weibull] as [Assumed distribution].
- Click [OK].

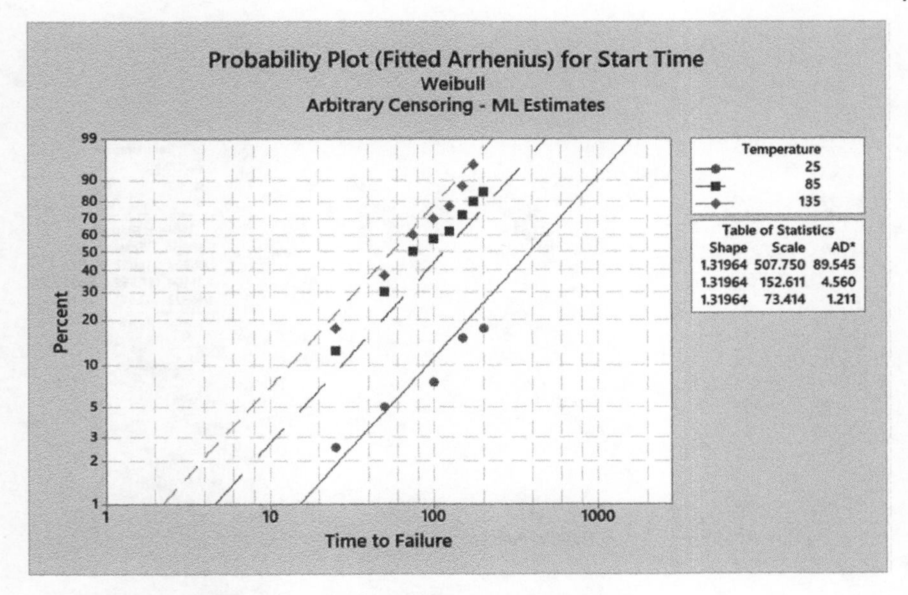

Figure 5.8 Probability plots of accelerated testing using Minitab.

The probability plot can be constructed as seen in Figure 5.8.

In Figure 5.8, the same shape parameter ($\beta = 1.31964$) was applied to all stress conditions. Individual fits of each condition could be applied to see whether the common slope is appropriate for this data.

Figure 5.9 shows the probability plot with individual fits of each condition. The three slopes do not look parallel. In this case, it is assumed that the failure mechanisms could be changed across different temperature conditions.

5.4.2 Python Practice

The components' use condition temperature is 35 °C, and the stress condition temperature is 145 °C. The ΔH is known to be 1.0. Calculate the acceleration factor using the Arrhenius model. In this case, Python codes can be written as shown in Figure 5.10.

After running all the codes, the *AF* can be reported as an output (Figure 5.11).

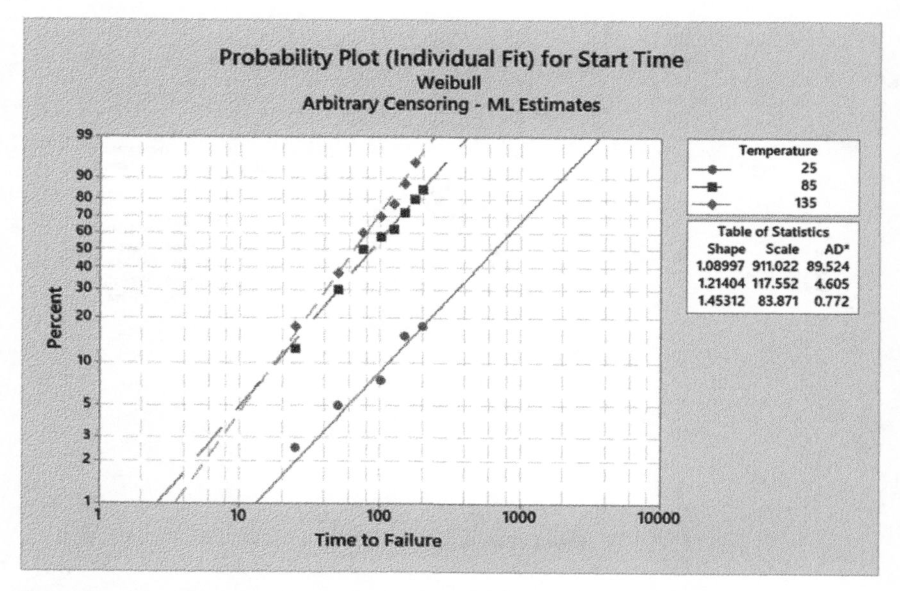

Figure 5.9 Probability plots with individual fit using Minitab.

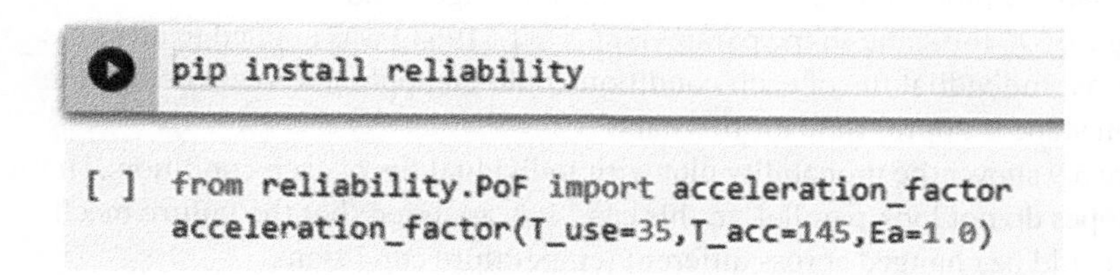

Figure 5.10 Python codes to calculate the *AF* in the Arrhenius model.

```
Results from acceleration_factor:
Acceleration Factor: 20062.75562462843
Use Temperature: 35 °C
Accelerated Temperature: 145 °C
Activation Energy: 1.0 eV
```

Figure 5.11 The *AF* in the Arrhenius model with Python.

5.5 Summary

- Based on the failure data from high-stress conditions, accelerated life testing projects the failure characteristics of components in normal or use conditions.
- In accelerated testing theory, it is assumed that the same failure mechanism holds through various stress levels. The failure time is only expected to be accelerated.
- The accelerated testing theory can be applied to various life distributions.
- The Arrhenius model can be used when failures are accelerated based on an increase in temperature.

Exercises

1. Describe the assumptions of accelerated life testing theory.

2. The semiconductor's mean time to failure (MTTF) at the use condition is 5 years. In the high-stress condition with a raised temperature, the MTTF was 1000 hours. The semiconductor was operated 10 hours per day. Calculate the acceleration factor.

3. A component was tested at 135 °C as a stress condition, and the MTTF was 3000 hours. The failure data was fitted to the exponential distribution. In the typical use temperature of 25 °C, calculate the probability of failure by 30,000 hours. The acceleration factor is known to be 40.

4. Based on the description in Exercise 3, determine the reliability at 40,000 hours at the use condition.

5. Diodes were tested under the elevated temperature of 145 °C. The failure data was related to the Weibull distribution with $\alpha = 2000$ hours and $\beta = 0.5$. An acceleration factor was known to be 50. Calculate the failure rate at 30,000 hours at the use condition.

6. Based on the description in Exercise 5, determine the reliability at 45,000 hours at the use condition.

7. The table describes the failure data with three different temperature conditions. Exact failure data of 10 components was collected for each condition,

and the testing was ended after 2000 hours. The failure data is fitted to the Weibull distribution. Use Minitab to construct a probability plot, and estimate the parameters of the Weibull distribution. Calculate the probability of failure by 3000 hours at the use condition.

Failure Order	25 °C	75 °C	135 °C
1	500	200	50
2	600	250	75
3	700	300	80
4	900	350	90
5	1300	450	120
6	1500	560	135
7	2000	1300	150
8		1400	170
9			300
10			400

8 Based on the description in Exercise 7, apply the Arrhenius model and compare the results.

6

System Failure Modeling

Chapter Overview and Learning Objectives

- To understand the objectives and characteristics of using a reliability block diagram.
- To understand the characteristics of the series system model, and calculate the system reliability.
- To learn the characteristics of the parallel system model, and draw a reliability block diagram.
- To calculate the system reliability of the combined serial–parallel system models.
- To learn the requirements of the *k*-out-of-*n* system models, and calculate the system reliability.
- To understand the characteristics of the minimal paths and minimal cuts, and calculate the system reliability.

6.1 Reliability Block Diagram

In previous chapters, we have been focusing on the failure characteristics of individual items. In reality, many devices and products are constructed as a system. A single smart phone consists of several hundreds of components inside. Some components could have similar failure mechanisms, whereas others could have completely different failure characteristics. This complexity is a motivation for understanding system failure modeling.

By studying system failure modeling, we can explore several questions:

- How do different designs of the system affect the overall reliability?
- What are the benefits of redundant design?
- What is the proper design (parallel or series) of components to achieve the desired reliability?

Reliability Analysis Using MINITAB and Python, First Edition. Jaejin Hwang.
© 2023 John Wiley & Sons, Inc. Published 2023 by John Wiley & Sons, Inc.
Companion Website: www.wiley.com\go\Hwang\ReliabilityAnalysisUsingMinitabandPython

Reliability block diagrams are a graphical way to demonstrate the system. Components are connected in terms of reliability, which could often be different from how components are physically related. Figure 6.1 shows an example of the reliability block diagrams of a simplified computer system. If the power supply fails, it will affect the reliability of the computer. If one fan fails, the computer may still operate. Based on the reliability block diagram, we could understand the effect of each component's failure or success on the whole system's failure or success.

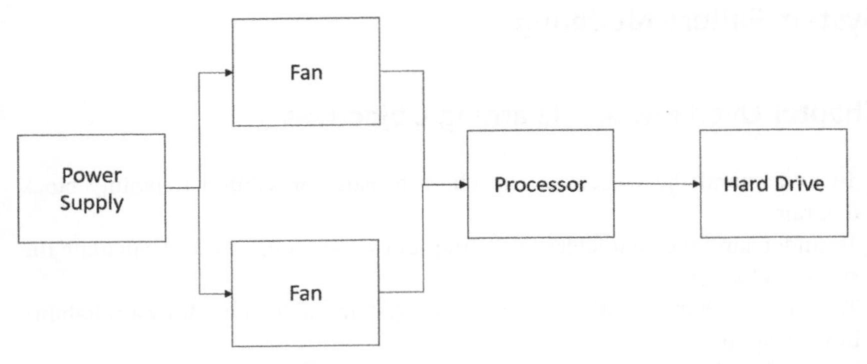

Figure 6.1 The reliability block diagram of a computer system.

6.2 Series System Model

The series system model is one of the most commonly used models in reliability. The system consists of independent components. If any component fails, the whole system also fails. The series system model is also called a first fail model or chain model (Figure 6.2).

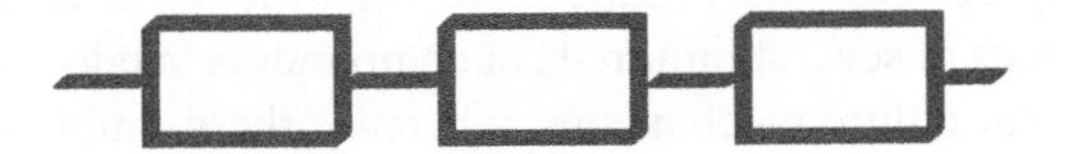

Figure 6.2 Series system model.

To calculate the system's reliability in the series system model, the joint probability can be considered. The joint probability is the probability of multiple events occurring simultaneously. If events are independent, the multiplication rule can be applied.

Given here is the equation to calculate the system reliability. This represents the probability that all components survive simultaneously by time *t*.

$$R_s(t) = \prod_{i=1}^{n} R_i(t) = R_1(t) \times R_2(t) \times \cdots \times R_n(t) \tag{6.1}$$

To understand the failure of the system, the union probability can be considered. The union probability is the probability that either of multiple events may occur.

Here is the equation to be driven to calculate the probability of failure of the system. If one of the components fails, it leads to the whole system's failure.

$$F_s(t) = F_1(t) + F_2(t) - F_1(t)F_2(t)$$

$$F_s(t) = 1 - (1 - F_1(t))(1 - F_2(t))$$

$$F_s(t) = 1 - \prod_{i=1}^{n} (1 - F_i(t)) \tag{6.2}$$

Example 6.1 In the series system model, each component has identical reliability of 0.90. Calculate the system reliability, as the number of components increases in the system (Table 6.1).

Table 6.1 The series system model's reliability by the number of components.

Number of Components	System Reliability
1	
2	
4	
6	
8	
10	

Answer:

As seen in the table, as the number of components increases, the system reliability is dramatically decreased. This indicates that each component should have high reliability, especially when the series system has many components.

Number of Components	System Reliability
1	0.9
2	0.81
4	0.6561
6	0.5314
8	0.4305
10	0.3487

Example 6.2 Three components are connected in a series of the system. The first component has a reliability of 0.90, the second component shows reliability of 0.95, and the third component has a reliability of 0.85 by 1000 hours. Calculate the overall reliability of the system by 1000 hours.

Answer:

$$R_s(1000) = R_1(1000) \times R_2(1000) \times R_3(1000) = 0.90 \times 0.95 \times 0.85 = 0.727$$

An exponential distribution can be applied to the series system model. The equation of the overall reliability of the system is

$$R_s(t) = \prod_{i=1}^{n} R_i(t) = \prod_{i=1}^{n} e^{-\lambda_i t} = e^{-\sum_{i=1}^{n} \lambda_i t} \tag{6.3}$$

Example 6.3 The series system model consists of three components. Each component's failure characteristics are related to the exponential distribution. The failure rate of the three components is 0.0005, 0.0001, and 0.0015. Calculate the reliability of the system by 1000 hours.

Answer:

$$R_s(1000) = e^{-\sum_{i=1}^{n} \lambda_i t} = e^{-(0.0005 + 0.0001 + 0.0015) \times 1000} = 0.1225$$

Example 6.4 The series system model consists of four components. Each component is independent and has an identical constant failure rate. The reliability of the system after 1000 hours is 0.90. Calculate the mean time to failure of an individual component.

Answer:

$$R_s(1000) = e^{-\sum_{i=1}^{n} \lambda_i t} = e^{-(4 \times \lambda \times 1000)} = 0.90$$

$$\lambda = \frac{-\ln 0.90}{4000} = 0.0000263$$

$$\text{MTTF} = \frac{1}{0.0000263} = 38022.8 \text{ hours}$$

The Weibull distribution can also be applied to the series system model. The equation of the system reliability is as follows:

$$R_s(t) = \prod_{i=1}^{n} e^{-\left(\frac{t}{\alpha_i}\right)^{\beta_i}} = e^{-\sum_{i=1}^{n}\left(\frac{t}{\alpha_i}\right)^{\beta_i}} \tag{6.4}$$

Example 6.5 The vehicle's axle system consists of left and right axles. The failure characteristics of these axles are related to the Weibull distribution with a shape parameter of 2.0. The characteristic life of each axle is 5000 and 6000 days. Calculate the reliability of the axle system by 3000 days.

Answer:

$$R_s(3000) = e^{-\sum_{i=1}^{n}\left(\frac{t}{\alpha_i}\right)^{\beta_i}} = e^{-\left(\frac{3000}{5000}\right)^{2.0}} \times e^{-\left(\frac{3000}{6000}\right)^{2.0}} = 0.5434$$

The lognormal distribution and normal distribution can also be applied to the series system model. The approach is similar to how exponential and Weibull distributions were applied.

6.3 Parallel System Model

The parallel system model is the opposite of the series system model (Figure 6.3). In a parallel system the system survives until the last component fails. The parallel system model is commonly applied in the aerospace industry. The support cables in bridges are another example of the parallel system model.

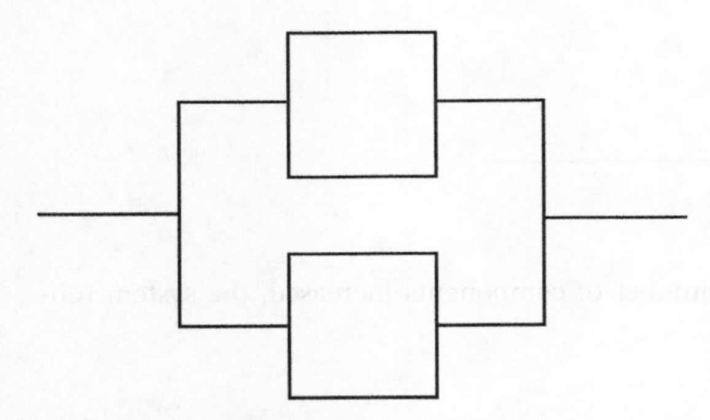

Figure 6.3 Parallel system model.

The parallel system has advantages in maintaining reliability. For instance, computer systems in the space shuttle are replicated with several backup copies. However, there is a trade-off between cost and reliability. If the number of parallel components increased, the cost would increase as well.

In the parallel system model, there are two different types of redundancy.

Active redundancy represents a standard parallel system. All components in the system operate simultaneously. Passive or standby redundancy means that some components are inactive during the system operation. These parts will be operated only when existing active parts fail. A backup standby generator in a building could be an example.

We will focus on active redundancy in this book. The joint probability can be applied to calculate the probability of failure. This means that the system would only fail when all components fail.

$$F_S(t) = \prod_{i=1}^{n} F_i(t) = F_1(t) \times F_2(t) \times \cdots \times F_n(t) \tag{6.5}$$

The union probability can be applied to calculate the system reliability.

$$R_S(t) = 1 - \prod_{i=1}^{n} \left(1 - R_i(t)\right) \tag{6.6}$$

Example 6.6 In the parallel system model, each component has identical reliability of 0.90. Calculate the system reliability, as the number of components in the system increases (Table 6.2).

Table 6.2 The parallel system model's reliability by the number of components.

Number of Components	System Reliability
1	
2	
4	
6	

Answer:

As seen in the table, as the number of components increased, the system reliability increased as well.

Number of Components	System Reliability
1	0.9
2	0.99
4	0.9999
6	0.999999

Example 6.7 Three components are connected in parallel in the system. The first component has a reliability of 0.90, the second component shows reliability of 0.95, and the third component has a reliability of 0.85, each by 1000 hours. Calculate the overall reliability of the system by 1000 hours.

Answer:

$$R_s(1000) = 1 - (1 - 0.90)(1 - 0.95)(1 - 0.85) = 0.99925$$

The exponential distribution can be applied to the parallel system model. If all components exhibit constant failure rates, the overall reliability of the system can be calculated as

$$R_s(t) = 1 - \prod_{i=1}^{n} \left[1 - e^{-\lambda_i t}\right] \tag{6.7}$$

Example 6.8 The parallel system model consists of three components. Each component's failure characteristics are related to the exponential distribution. The failure rate of each component is 0.0005, 0.0001, and 0.0015. Calculate the reliability of the system by 1000 hours.

Answer:

$$R_s(1000) = 1 - \left[\begin{array}{l}\left(1 - e^{-(0.0005)\times(1000)}\right)\left(1 - e^{-(0.0001)\times(1000)}\right) \\ \left(1 - e^{-(0.0015)\times(1000)}\right)\end{array}\right] = 0.9709$$

The Weibull distribution can also be applied to the parallel system model. If each component of the system shows failure characteristics related to the Weibull distribution, the system reliability can be calculated as

$$R_s(t) = 1 - \prod_{i=1}^{n} \left[1 - e^{-\left(\frac{t}{\alpha_i}\right)^{\beta_i}}\right] \tag{6.8}$$

Example 6.9 A light fixture consists of three identical light bulbs that are connected in parallel. Each bulb exhibits a Weibull failure distribution with $\beta = 0.75$, and a characteristic life is 10 months. Calculate the reliability of a light fixture by 1 year.

Answer:

$$R_s(12) = 1 - \left[1 - e^{-\left(\frac{12}{10}\right)^{0.75}}\right]\left[1 - e^{-\left(\frac{12}{10}\right)^{0.75}}\right]\left[1 - e^{-\left(\frac{12}{10}\right)^{0.75}}\right] = 0.6824$$

Example 6.10 Calculate the minimum number of redundant light bulbs to achieve system reliability of 0.90. Each bulb has a reliability of 0.6 and they are connected in parallel.

Answer:

$$R_s(t) = 1 - (1 - 0.6)^n = 0.90$$

$$0.4^n = 0.1$$

$$\text{n} = \frac{\ln 0.1}{\ln 0.4} = 2.51 => 3 \text{ bulbs}$$

6.4 Combined Serial–Parallel System Model

We have discussed the series and parallel system models. A system often contains components connected in both series and parallel. In this case, the system could be divided into several subsystems to calculate the system reliability.

Example 6.11 As seen in Figure 6.4, the system consists of three components connected both in series and in parallel. The reliability by 1000 hours of the three components is 0.9, 0.95, and 0.9. Calculate the system reliability by 1000 hours.

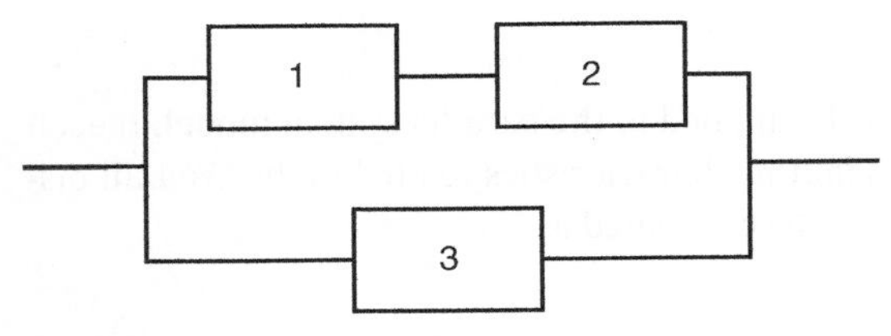

Figure 6.4 Combined serial–parallel system model with three components.

Answer:

For components 1 and 2

$$R_s(1000) = 0.9 \times 0.95 = 0.855$$

$$R_s(1000) = 1 - (1 - 0.855)(1 - 0.9) = 0.9855$$

Example 6.12 As seen in Figure 6.5, the system consists of components connected in both series and parallel. Each component's reliability by 10 months is shown in the figure. Calculate the overall system reliability.

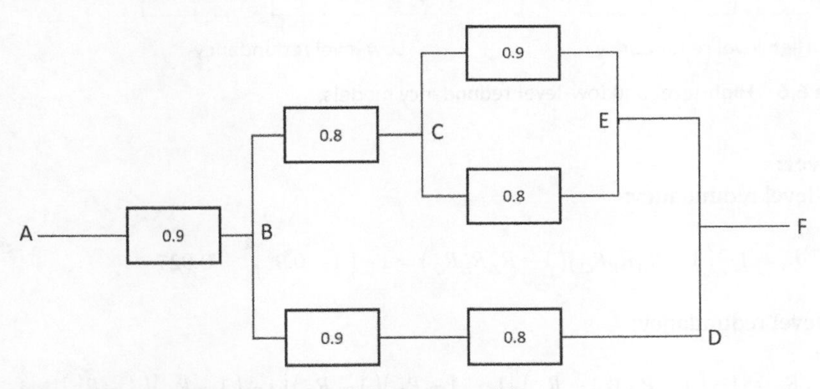

Figure 6.5 Combined serial–parallel system model.

Answer:

$$R_{C-E}(10) = 1 - (1 - 0.9)(1 - 0.8) = 0.98$$

$$R_{B-C-E}(10) = 0.8 \times 0.98 = 0.784$$

$$R_{B-D}(10) = 0.9 \times 0.8 = 0.72$$

$$R_{B-F}(10) = 1 - (1 - 0.784)(1 - 0.72) = 0.93952$$

$$R_{A-F}(10) = 0.9 \times 0.93952 = 0.8456$$

The combined system has two different options, including high-level redundancy and low-level redundancy. For high-level redundancy, the entire system could be connected in parallel. For low-level redundancy, each component could be connected in parallel.

Example 6.13 A company has two choices of constructing a combined system (Figure 6.6). The system would consist of three components, and these components could be in parallel with either high-level redundancy or low-level redundancy. Each component has the same reliability of 0.90. Calculate the system reliability of each case, and choose the system that could lead to higher system reliability.

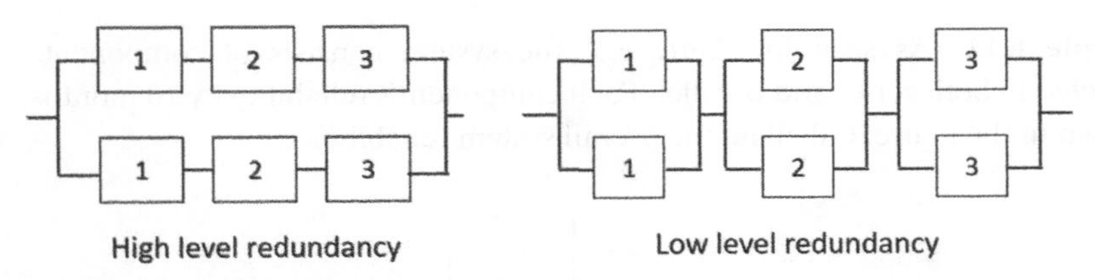

Figure 6.6 High-level and low-level redundancy models.

Answer:
High-level redundancy:

$$R_s = 1-(1-R_A R_B R_C)(1-R_A R_B R_C) = 1-\left(1-0.9^3\right)^2 = 0.927$$

Low-level redundancy:

$$\begin{aligned} R_s &= [1-(1-R_A)(1-R_A)][1-(1-R_B)(1-R_B)][1-(1-R_C)(1-R_C)] \\ &= \left[1-(1-0.9)^2\right]^3 \\ &= 0.970 \end{aligned}$$

The low-level redundancy system shows higher reliability than the high-level redundancy system.

6.5 *k*-out-of-*n* System Model

The *k*-out-of-*n* system model is a special case of the series and parallel systems. The system requires that at least *k* components should survive out of *n* components. For the extreme case, if *k* is 1, the system is the same as a parallel system model. If *k* is *n*, it is identical to the series system model. For instance, an airplane includes four engines. If at least two engines are required to operate for the aircraft's survival, it could be written as a 2-out-of-4 system.

The reliability of the system can be calculated using the binomial formula.

$$R_s(t) = \sum_{i=k}^{n} \binom{n}{i} R(t)^i F(t)^{n-i}$$

$$\binom{n}{i} = \frac{!}{i!(n-i)!} \tag{6.9}$$

For example, for the 2-out-of-4 system, the reliability can be written as

$$R_s(t) = \sum_{i=2}^{4} \binom{4}{i} R(t)^i F(t)^{4-i}$$
$$= \binom{4}{2} R(t)^2 F(t)^2 + \binom{4}{3} R(t)^3 F(t)^1 + \binom{4}{4} R(t)^4 F(t)^0 \tag{6.10}$$

Example 6.14 In a robot assembly, the system includes three identical robot arms. To maintain the operation of the assembly line, at least two robot arms should function. Each robot arm's reliability by 1000 hours is 0.90. Calculate the system reliability by 1000 hours.

Answer:
The reliability block diagram (RBD) can be illustrated as in Figure 6.7.

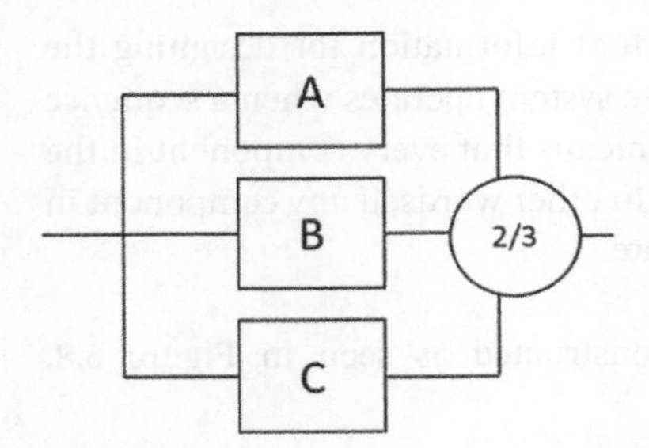

Figure 6.7 2-out-of-3 system model.

$$R_s(1000) = \sum_{i=2}^{3} \binom{3}{i} R(t)^i F(t)^{3-i}$$
$$= \binom{3}{2} R(t)^2 F(t)^1 + \binom{3}{3} R(t)^3 F(t)^0 = \binom{3}{2} 0.9^2 0.1^1 + \binom{3}{3} 0.9^3 0.1^0$$
$$= 0.972$$

The exponential distribution can be applied to the k-out-of-n system model. The system reliability can be calculated as

$$R_s(t) = \sum_{i=k}^{n} \binom{n}{i} e^{-\lambda i t} \left[1 - e^{-\lambda t}\right]^{n-i} \tag{6.11}$$

Example 6.15 A web server system consists of four identical servers. At least three of them should function to make the web server system operational. Each server's failure characteristics are based on the exponential distribution with $\lambda = 0.000003$ per hour. Calculate the system's reliability by 10,000 hours.

Answer:

$$R_s(10000) = \sum_{i=3}^{4} \binom{4}{i} e^{-0.000003 \times i \times 10000} \left[1 - e^{-0.000003 \times 10000}\right]^{4-i}$$

$$= \binom{4}{3} e^{-0.000003 \times 3 \times 10000} \left[1 - e^{-0.000003 \times 10000}\right]^{1}$$

$$+ \binom{4}{4} e^{-0.000003 \times 4 \times 10000} \left[1 - e^{-0.000003 \times 10000}\right]^{0} = 0.995$$

6.6 Minimal Paths and Minimal Cuts

Minimal paths and minimal cuts are important information for designing the structure of the system. A path means that the system operates when a sequence of all components operates. A minimal path means that every component in the path is critical to the operation of the system. In other words, if any component in the path fails, the system will no longer operate.

Example 6.16 A bridge structure was constructed as seen in Figure 6.8. Determine the minimal paths of the system.

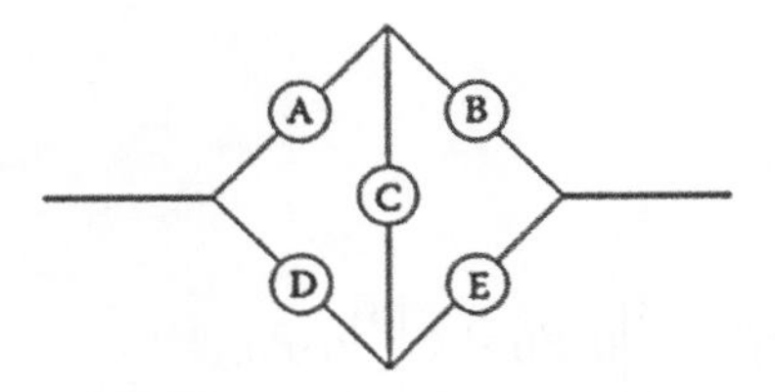

Figure 6.8 A bridge structure.

Answer:
The minimal paths are as follows:

$$(A, B), (A, C, E), (D, E), (D, C, B)$$

Example 6.17 Based on the bridge structure of Example 6.16, calculate the system reliability using minimal paths.

Answer:
The RBD can be drawn as in Figure 6.9.

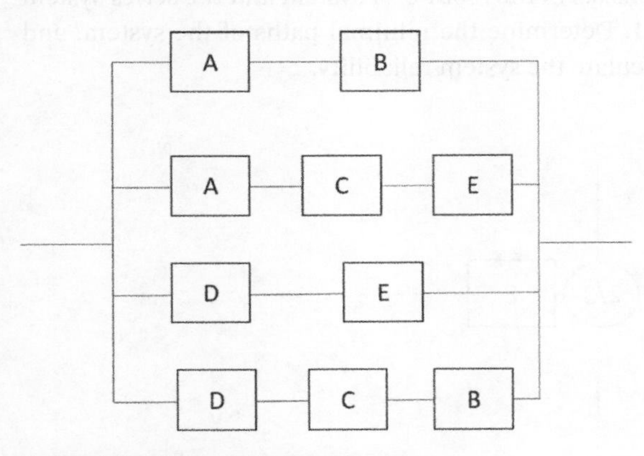

Figure 6.9 RBD of a bridge structure.

The system reliability can be calculated as follows:

$$R_s(t) = 1-(1-R_A(t)R_B(t))(1-R_A(t)R_C(t)R_E(t))$$
$$(1-R_D(t)R_E(t))(1-R_D(t)R_C(t)R_B(t))$$

Example 6.18 Three components are connected in series and in parallel as seen in Figure 6.10. Identify the minimal paths and calculate the system reliability.

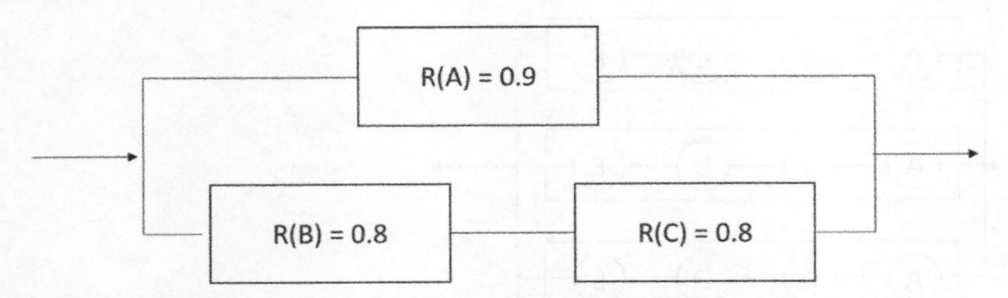

Figure 6.10 Combined serial–parallel system model with three components.

Answer:
The minimal paths are as follows:

$$(A), (B, C)$$

The system reliability can be calculated as

$$R_s = 1 - \left[(1-0.9)(1-0.8\times0.8)\right] = 0.964$$

Example 6.19 A system consists of the *k*-out-of-*n* system and the series system model as seen in Figure 6.11. Determine the minimal paths of the system, and construct the equation to calculate the system reliability.

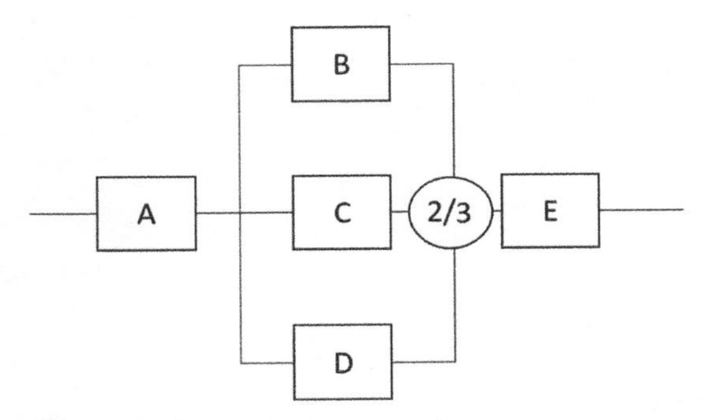

Figure 6.11 Combined serial–parallel system model including *k*-out-of-*n* system model.

Answer:
The minimal paths are as follows:

$$(A, B, C, E), (A, B, D, E), (A, C, D, E)$$

The RBD can be drawn (Figure 6.12).
The system reliability can be calculated:

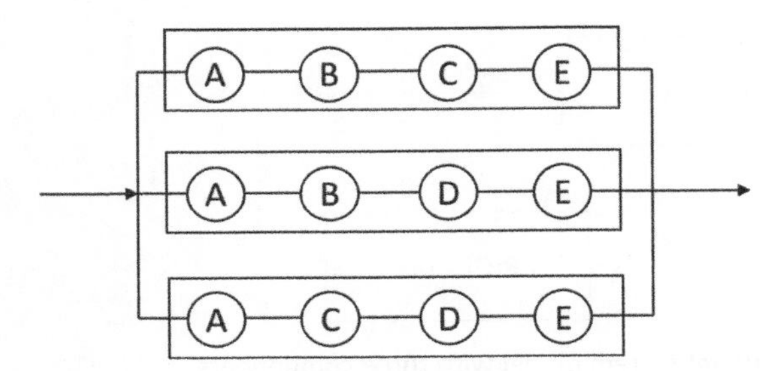

Figure 6.12 RBD of the minimal paths.

$$R_s = 1-\left(1-R_A R_B R_C R_E\right)\left(1-R_A R_B R_D R_E\right)\left(1-R_A R_C R_D R_E\right)$$

A cut is the opposite concept of a path. A cut means that a failure of the component in the sequence would result in system failure. A minimal cut means that all components in a sequence must fail to make the system fail. Information on minimal cut sets can be helpful to understand the structural vulnerability of the system. If the system consists of numerous cut sets, the system will exhibit high vulnerability. In addition, if there is a single point failure in the system, this could lead to high vulnerability of the system. Minimal cut set information can help to diagnose this vulnerability.

The fault tree diagram is a concept opposite to the reliability block diagram (RBD). While the RBD considers the survival or reliability of a system, the fault tree diagram addresses the failure characteristics of the system. The fault tree diagram could be closely related to the minimal cut sets.

Example 6.20 A system is constructed with series and parallel system models as seen in Figure 6.13. Determine the minimal cuts and construct the fault tree diagram. Convert the fault tree diagram to the RBD.

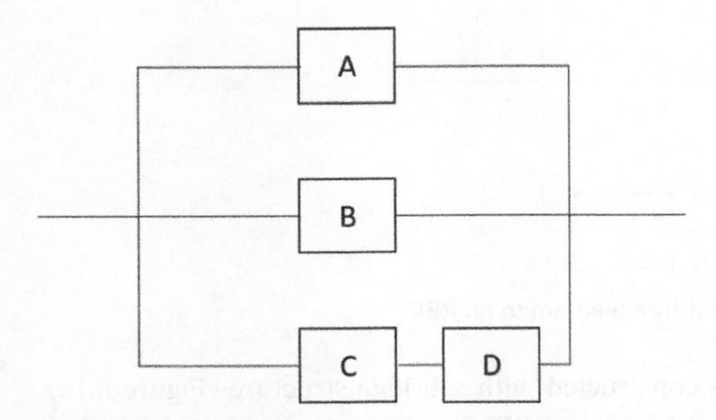

Figure 6.13 Combined serial–parallel system model with four components.

Answer:

The minimal cuts are as follows:

$$(A, B, C), (A, B, D)$$

The fault tree diagram can be constructed as seen in Figure 6.14.

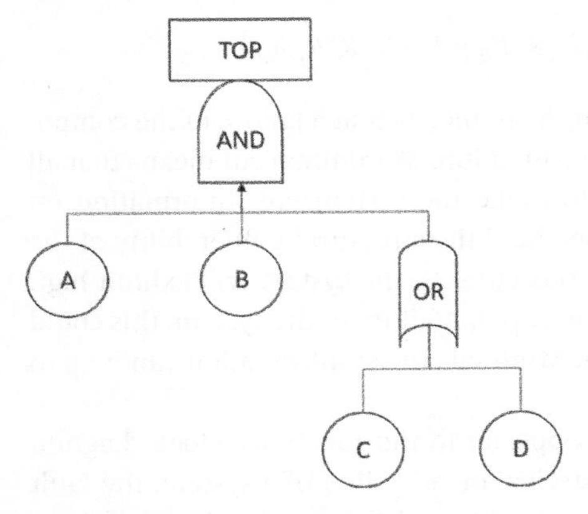

Figure 6.14 The fault tree diagram.

The fault tree diagram can be converted to an RBD as seen in Figure 6.15.

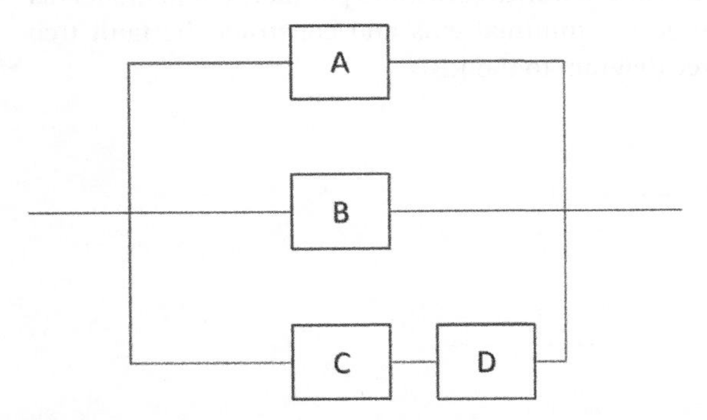

Figure 6.15 Converting the fault tree diagram to an RBD.

Example 6.21 A system is constructed with a bridge structure (Figure 6.16). Determine the minimal cuts and draw an RBD. Construct an equation to calculate the system reliability.

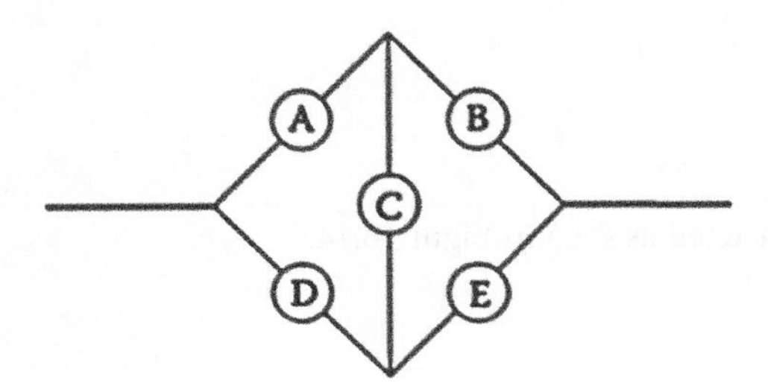

Figure 6.16 The bridge structure with five components.

Answer:
The minimal cuts are as follows:

$$(A, D), (B, E), (A, C, E), (D, C, B)$$

The RBD can be drawn as seen in Figure 6.17.

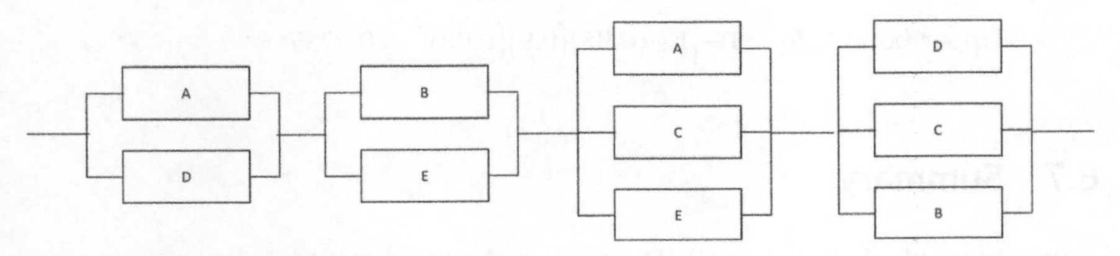

Figure 6.17 The RBD of minimal cuts.

The system reliability can be calculated as

$$R_s(t) = [1-(1-R_A)(1-R_D)][1-(1-R_B)(1-R_E)]$$
$$[1-(1-R_A)(1-R_C)(1-R_E)][1-(1-R_D)(1-R_C)(1-R_B)]$$

Based on the information of minimal paths and minimal cuts, the upper and lower bounds of the system can be determined. The upper bound of the reliability can be determined by considering the minimal path sets. The lower bound of the reliability can be calculated using the minimal cut sets.

Example 6.22 Identify the minimal path sets and minimal cut sets (Figure 6.18). Calculate the lower and upper bounds of reliabilities.

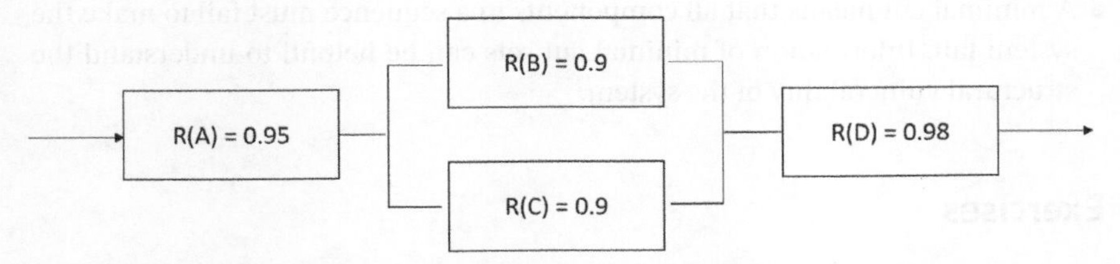

Figure 6.18 Combined serial–parallel system model with four components.

Answer:
The minimal paths are as follows:

$$(A, B, D), (A, C, D)$$

The minimal cuts are

$$(A), (D), (B, C)$$

The lower bound can be calculated using minimal cut sets.

$$\text{Lower bounds } R_s = (0.95)(0.98)[1-(1-0.9(1-0.9)] = 0.92169$$

The upper bound can be calculated using minimal path sets.

$$\text{Upper bounds } R_s = 1-[1-(0.95)(0.9)(0.98)]^2 = 0.9737$$

6.7 Summary

- Reliability block diagrams (RBD) are a graphical way to demonstrate a system. Components are connected in terms of reliability.
- The series system model consists of independent components. If any component fails, the whole system also fails. The series system model is also called a first fail model or chain model.
- The parallel system model is the opposite of the series system model. In the parallel system model, a system survives until the last component fails.
- A system often contains components connected in both series and parallel. In this case, the system could be divided into several subsystems to calculate the system reliability.
- The *k*-out-of-*n* system model is a special case of the series and parallel systems. The system requires that at least *k* components should survive out of *n* components.
- A minimal path means that every component in the path is critical to the operation of the system. In other words, if any component in the path fails, the system will no longer operate.
- A minimal cut means that all components in a sequence must fail to make the system fail. Information of minimal cut sets can be helpful to understand the structural vulnerability of the system.

Exercises

1 The following system consists of the series and parallel system models. Each component has the same constant failure rate. Determine each component's mean time to failure to reach the system reliability of 0.80 by 80 hours.

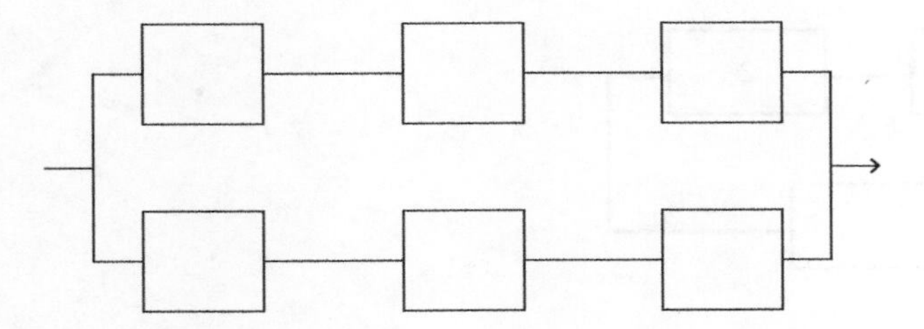

2 Calculate the system reliability of the following system. The reliability of component 1, 2, and 3 is 0.95, 0.97, and 0.96, respectively.

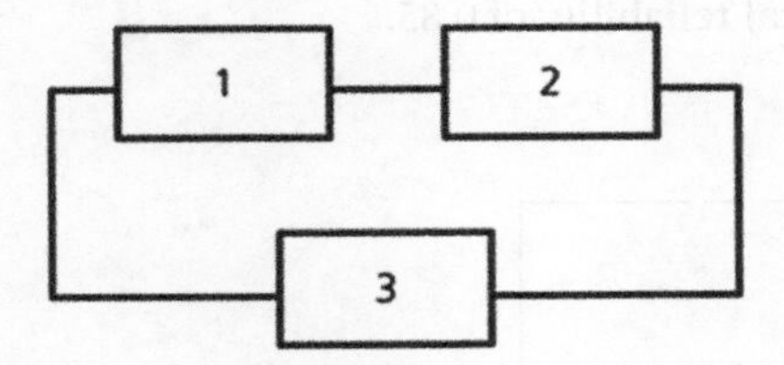

3 A pump system includes three identical pumps. At least two pumps should operate properly to make the whole system function. Each pump's reliability by 1000 hours is 0.90. Calculate the system reliability by 1000 hours.

4 Describe the differences between high-level redundancy and low-level redundancy models.

5 Determine the minimal cuts of the following system. Draw a fault tree diagram.

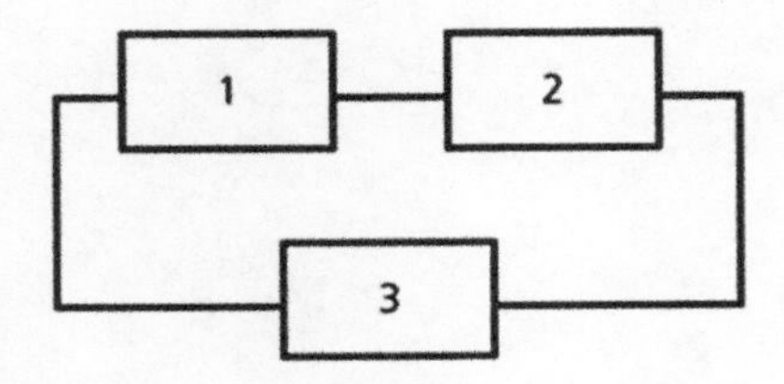

6 Determine the minimal paths of the following system. Draw a reliability block diagram.

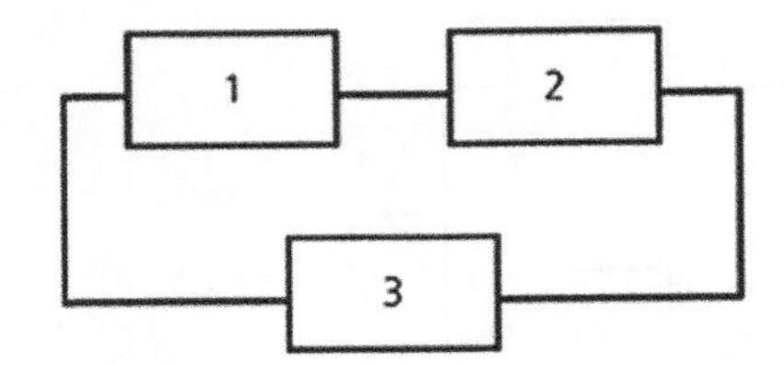

7 Discuss the differences between the reliability block diagram and the fault tree diagram.

8 Calculate the lower bound and upper bound of the reliability in the following system. Each component has identical reliability of 0.85.

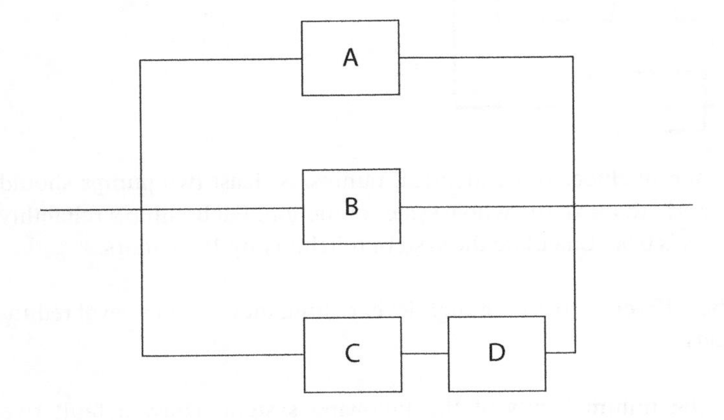

7

Repairable Systems

Chapter Overview and Learning Objectives

- To understand the objective and definition of corrective maintenance and preventive maintenance.
- To learn a measure of repairable systems, such as the mean time between failure and mean time to repair.
- To understand the concept of availability and differentiate different availability measures, including inherence availability, achieved availability, and operational availability.
- To understand the fundamental concept of maintainability.
- To determine the optimal preventive maintenance schedule to minimize the total cost.

7.1 Corrective Maintenance

In previous chapters, we assumed that failed products were discarded and were not repaired to bring them back to their original condition. However, in real life, components are often repaired and returned to operation. In that case, the maintenance and availability of the components and systems are important concerns. We can explore several questions:

- How long does it take for the component to be repaired?
- What are the repair characteristics of the components (a repair probability distribution)?
- How often should we perform maintenance to maintain the desired reliability of the components or systems?

Reliability Analysis Using MINITAB and Python, First Edition. Jaejin Hwang.
© 2023 John Wiley & Sons, Inc. Published 2023 by John Wiley & Sons, Inc.
Companion Website: www.wiley.com\go\Hwang\ReliabilityAnalysisUsingMinitabandPython

We could initially discuss maintenance. Maintenance can be categorized into corrective maintenance and preventive maintenance.

Corrective maintenance focuses on correcting the failures of the components. It is considered a reactive process because it is only conducted when failure is observed. This means that corrective maintenance is performed at unpredictable intervals since the occurrence of failure is often difficult to predict. As a quantitative measure, the mean time to repair (MTTR) could be calculated. The maintenance goal is to minimize the MTTR and restore the component and system to a satisfactory condition.

7.2 Preventive Maintenance

Preventive maintenance is the opposite of corrective maintenance and is an activity to replace or repair components before they fail. The goal is to maintain the system operation in a satisfactory manner without pausing the operation due to failures. For instance, servicing (cleaning and lubrication) and inspection activities could be performed. Scheduling is an important matter in preventive maintenance since it is critically related to the cost. Based on the historical failure or repair data of the components, the ideal intervals of preventive maintenance could be determined.

7.3 Mean Time between Failures

The mean time between failures (MTBF) is a measure to quantify the amount of operation times between failures. The design goal is to increase the MTBF as much as possible. Here is the formula to calculate the MTBF:

$$\text{MTBF} = \frac{\text{Operating time}}{\text{Number of failures}} \tag{7.1}$$

Figure 7.1 illustrates examples of failure and repair in a system.

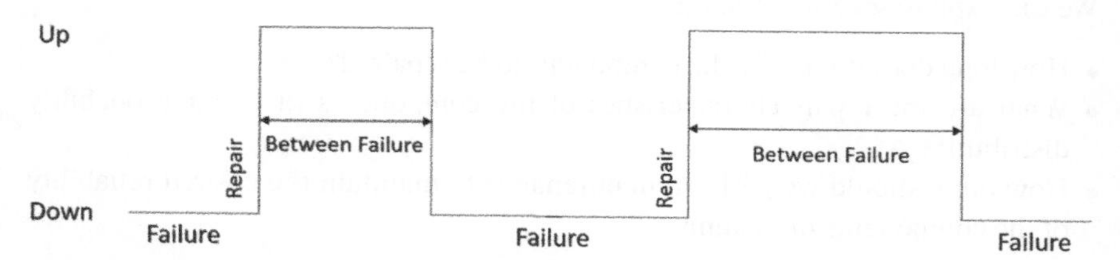

Figure 7.1 Illustration of the mean time between failures.

7.4 Mean Time to Repair

The mean time to repair (MTTR) is another measure to evaluate the efficiency of corrective maintenance. The design goal is to minimize the MTTR to return the system to its original condition as soon as possible. The equation of the MTTR is as follows:

$$\text{MTTR} = \frac{\text{Downtime}}{\text{Number of failures}} \tag{7.2}$$

Example 7.1 Based on the system's operation described in Figure 7.2, calculate the MTBF and MTTR.

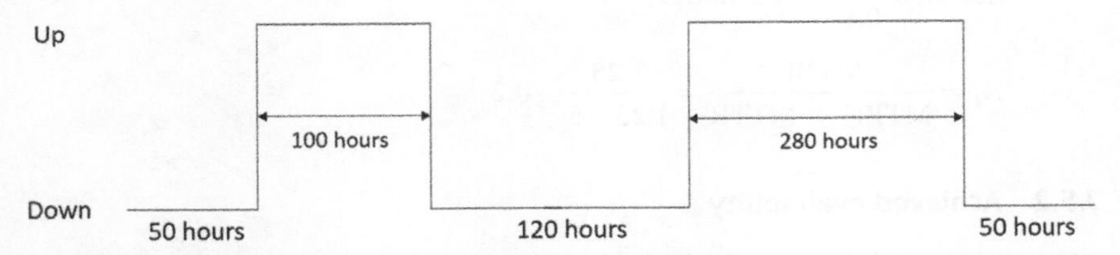

Figure 7.2 The illustration of the mean time between failure and mean time to repair.

Answer:

$$\text{MTBF} = \frac{\text{Operating time}}{\text{Number of failures}} = \frac{100 + 280}{3} = 126.67 \text{ hours}$$

$$\text{MTTR} = \frac{\text{Downtime}}{\text{Number of failures}} = \frac{50 + 120 + 50}{3} = 73.33 \text{ hours}$$

7.5 Availability

Availability is the probability that a system is in operation when requested for use. The availability could be affected by the reliability and maintainability of the system. The availability can be calculated as:

$$\text{Availability} = \frac{\text{Uptime}}{\text{Uptime} + \text{Downtime}} \tag{7.3}$$

Depending on the types of downtime, there are some variations of availability measures.

7.5.1 Inherent Availability

Inherent availability only considers corrective maintenance as a downtime, which means that it includes preventive maintenance and other administrative downtimes when calculating availability. The equation of the inherent availability is as follows:

$$A_I = \frac{\text{MTBF}}{\text{MTBF} + \text{MTTR}} \tag{7.4}$$

Example 7.2 A computer server's time between failures is related to the exponential distribution with $\lambda = 0.8$. The average time to repair the server is 5 hours. Calculate the inherent availability of the computer server.

Answer:

$$\text{MTBF} = \frac{1}{0.8} = 1.25 \text{ hours}$$

$$A_I = \frac{\text{MTBF}}{\text{MTBF} + \text{MTTR}} = \frac{1.25}{1.25 + 5} = 0.2$$

7.5.2 Achieved Availability

Achieved availability considers both corrective maintenance and preventive maintenance as downtimes. It does not consider administrative downtime. The equation for achieved availability is as follows:

$$A_A = \frac{\text{MTBMA}}{\text{MTBMA} + \text{MMT}} \tag{7.5}$$

MTBMA is the mean time between any maintenance activities. MMT is the mean maintenance time by considering both preventive and corrective maintenance.

If we assume the constant failure rate of the system, the mean time between maintenance (MTBMA) can be calculated as

$$\text{MTBMA} = \frac{\text{Uptime}}{\text{Number of maintenance activities}} = \frac{1}{\lambda + f_{PM}} \tag{7.6}$$

The λ is a failure rate, and it can be directly related to the corrective maintenance. The f_{PM} is the number of preventive maintenance activities.

Mean maintenance time (MMT) can be calculated as

$$\text{MMT} = \frac{\lambda \text{MTTR} + f_{PM} \text{MPMT}}{\lambda + f_{PM}} \tag{7.7}$$

MTTR is the mean time to repair, and it could be considered as a mean corrective maintenance time. MPMT is the mean preventive maintenance time.

Example 7.3 A system shows a constant failure rate of 1 failure per 300 hours of operation. The average corrective maintenance duration is 8 hours. Preventive maintenance is conducted every 600 hours of operation. The average preventive maintenance duration is 5 hours. Calculate the achieved availability of the system.

Answer:

$$\lambda = \frac{1}{300} = 0.0033$$

$$f_{PM} = \frac{1}{600} = 0.0017$$

$$\text{MTBMA} = \frac{1}{\lambda + f_{PM}} = \frac{1}{0.0033 + 0.0017} = 200 \text{ hours}$$

$$\text{MMT} = \frac{0.0033 \times 8 + 0.0017 \times 5}{0.0033 + 0.0017} = 6.98 \text{ hours}$$

$$A_A = \frac{\text{MTBMA}}{\text{MTBMA} + \text{MMT}} = \frac{200}{200 + 6.98} = 0.966$$

7.5.3 Operational Availability

Operational availability considers all experienced sources of downtime, such as corrective and preventive maintenance time, and administrative downtime. This availability could be closely related to what a customer actually experiences. The equation of the operational availability is

$$A_{Op} = \frac{\text{MTBMA}}{\text{MTBMA} + \text{MDT}} \tag{7.8}$$

MTBMA is the mean time between maintenance activities. MDT is a mean downtime, and it could consist of all types of downtime.

Example 7.4 A system shows the constant failure rate of 1 failure per 500 hours of operation. The average corrective maintenance duration is 15 hours. Preventive maintenance is conducted every 300 hours of operation. The average preventive maintenance duration is 8 hours. The logistics and administrative delay time is 10 hours. Calculate the operational availability of the system.

Answer:

$$\text{MTBMA} = \frac{1}{\lambda + f_{PM}} = \frac{1}{0.002 + 0.0033} = 188.68 \text{ hours}$$

$$\text{MMT} = \frac{\lambda \text{MTTR} + f_{PM}\text{MPMT}}{\lambda + f_{PM}} = \frac{0.002 \times 15 + 0.0033 \times 8}{0.002 + 0.0033} = 10.64 \text{ hours}$$

$$\text{MDT} = \text{MMT} + \text{Delay time} = 10.64 + 10 = 20.64 \text{ hours}$$

$$A_{Op} = \frac{\text{MTBMA}}{\text{MTBMA} + \text{MDT}} = \frac{188.68}{188.68 + 20.64} = 0.901$$

7.5.4 System Availability

Since availability is a probability, the system availability can be calculated by considering the series and parallel models. The equation of the availability of the system in series is as follows:

$$A_s(t) = \prod_{i=1}^{n} A_i(t) \tag{7.9}$$

The equation of the availability of the system in parallel is

$$A_s(t) = 1 - \prod_{i=1}^{n} \left(1 - A_i(t)\right) \tag{7.10}$$

Example 7.5 A system consists of 2 components in series. Each component has the MTBF as 1000 hours and MTTR as 5 hours. Calculate the system's inherent availability.

Answer:

For each component,

$$A_I = \frac{\text{MTBF}}{\text{MTBF} + \text{MTTR}} = \frac{1000}{1000 + 5} = 0.995$$

$$A_s = A_1 A_2 = 0.995^2 = 0.99$$

7.6 Maintainability

Maintainability is the probability of repair in a specific time period. To understand the performance of maintainability, the mean time to repair (MTTR) and maintenance hours per operating hour (MH/OH) could be considered.

If we assume the constant failure rate of the system, the MH/OH can be calculated as

$$\frac{\text{MH}}{\text{OH}} = \frac{f_{CM} \times \text{MTTR} \times \text{CREW}}{t} \tag{7.11}$$

The f_{CM} is the number of failures. CREW is the average crew size needed to complete the repair.

If the MH/OH also includes the mean preventive maintenance time (MPMT), the MH/OH can be calculated as

$$\frac{\text{MH}}{\text{OH}} = \frac{f_{CM} \times \text{MTTR} \times \text{CREW} + f_{PM} \times \text{MPMT} \times \text{CREW}}{t} \tag{7.12}$$

The f_{PM} is the number of preventive maintenance actions.

Example 7.6 A system's repair time is normally distributed with a mean of 5 hours and a standard deviation of 1.5 hours. The average crew size necessary to repair the system is 2 members. Failure occurred every 200 hours. The design goal was that the MH/OH does not exceed 3 corrective maintenance hours for every 100 operating hours. Calculate the MH/OH and determine whether the design goal was met.

Answer:

$$\frac{\text{MH}}{\text{OH}} = \frac{f_{CM} \times \text{MTTR} \times \text{CREW}}{t} = \frac{1 \times 5 \times 2}{200} = 0.05$$

Since MH/OH is greater than 0.03, the design goal was not met.

Example 7.7 A system's repair time is normally distributed with a mean of 5 hours and a standard deviation of 1.5 hours. The average crew size necessary to repair the system is 2 members. Failure occurred every 200 hours. The preventive maintenance is conducted every 100 operating hours. The mean preventive maintenance time is 1 hour. The design goal was that the MH/OH does not exceed 3 corrective maintenance hours for every 100 operating hours. Calculate the MH/OH and determine whether the design goal was met.

Answer:

$$\frac{\text{MH}}{\text{OH}} = \frac{f_{CM} \times \text{MTTR} \times \text{CREW} + f_{PM} \times \text{MPMT} \times \text{CREW}}{t}$$

$$= \frac{1 \times 2 \times 2 + 2 \times 1 \times 2}{200} = 0.04$$

Since MH/OH is greater than 0.03, the design goal was not met.

7.7 Preventive Maintenance Scheduling

Preventive maintenance scheduling is critical to maximizing the effectiveness and economy of preventive maintenance activities. To accurately estimate the timing of preventive maintenance, understanding the failure distributions of the parts is necessary. In addition, the costs for the repair and downtime of the system should be calculated to make a better decision.

Figure 7.3 describes the reliability bathtub curve that we discussed in previous chapters. What would be the ideal time period to schedule preventive maintenance actions?

For the early life stage, the failure rate is dramatically decreasing over time. In this trend, if the replacement is conducted, the part would return to a high failure rate. Thus, there would be no benefit in applying preventive maintenance at this stage.

For the useful life stage, the constant failure rate is expected regardless of the time. This indicates that preventive maintenance actions would have no effect on changing the failure rate.

For the wearout life stage, the failure rate is increasing over time. In this trend, preventive maintenance would have a benefit in reducing the failure rate.

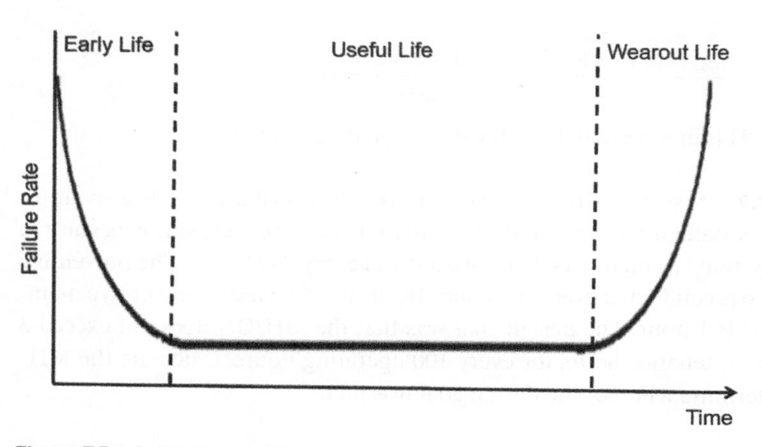

Figure 7.3 Reliability bathtub curve.

To identify the optimal timing of preventive maintenance, the total cost could be estimated. The following components could be considered to calculate the total cost:

- The cost of failure
- The cost of scheduled replacement
- The cost of inspection

Figure 7.4 illustrates the cost related to the maintenance interval. As the maintenance interval increases, the number of failures could increase as well. This could result in the rise of the corrective maintenance cost. As the maintenance interval decreases, more resources for preventive maintenance would be spent. The total

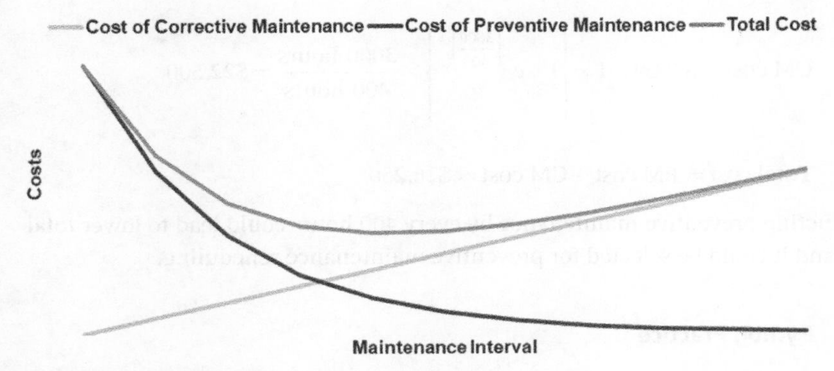

Figure 7.4 The cost related to the maintenance interval.

cost displays the U-shape. This suggests that it is important to find the optimal maintenance interval that could result in the minimized total cost.

Example 7.8 A system's failure characteristic is modeled by the Weibull distribution with $\beta = 1.5$ and $\alpha = 200$ hours. If failure occurs, the cost of repairing and downtime of the system is \$3000. For preventive maintenance, the cost of replacement is \$500. The system operates 3000 hours per year. The reliability engineer wants to determine whether preventive maintenance by every 200 hours or 400 hours is the more economic schedule. Calculate the total cost, and determine the schedule.

Answer:

Since the β is 1.5 of the Weibull distribution, the failure rate is expected to increase over time, so preventive maintenance activity could be effective in this case.

For preventive maintenance by every 200 hours:

$$\text{PM cost} = \frac{3000 \text{ hours}}{200 \text{ hours}} \times \$500 = \$7500$$

$$\text{CM cost} = \$3000 \times 1 \times \left\{1 - e^{-\left(\frac{3000}{200}\right)^{1.5}}\right\} \times \frac{3000 \text{ hours}}{200 \text{ hours}} = \$45{,}000$$

$$\text{Total cost} = \text{PM cost} + \text{CM cost} = \$52{,}500$$

For preventive maintenance by every 400 hours:

$$\text{PM cost} = \frac{3000 \text{ hours}}{400 \text{ hours}} \times \$500 = \$3750$$

$$\text{CM cost} = \$3000 \times 1 \times \left\{ 1 - e^{-\left(\frac{3000}{400}\right)^{1.5}} \right\} \times \frac{3000 \text{ hours}}{400 \text{ hours}} = \$22{,}500$$

$$\text{Total cost} = \text{PM cost} + \text{CM cost} = \$26{,}250$$

Conducting preventive maintenance by every 400 hours could lead to lower total cost, and it could be selected for preventive maintenance scheduling.

7.7.1 Python Practice

Python can be used to calculate the optimal scheduling of the replacement to minimize the total cost (Figure 7.5). For example, the cost of each preventive maintenance activity is $100. The cost of each corrective maintenance activity is $1000. The failure characteristics are modeled by the Weibull distribution with $\alpha = 300$ hours. $\beta = 1.5$.

[optimal_replacement_time] function can be imported.
[cost_PM, cost_CM, weibull_alpha, weibull_beta] can be set.

```
pip install reliability
```

```
[ ] pip install matplotlib==3.1.3
```

```
[ ] from reliability.Repairable_systems import optimal_replacement_time
```

```
[ ] import matplotlib.pyplot as plt
```

```
[ ] optimal_replacement_time(cost_PM=100, cost_CM=1000, weibull_alpha=300, weibull_beta=1.5,q=0)
    plt.show()
```

Figure 7.5 Python codes to determine the optical replacement interval.

After running all the codes, output results and associated plots were created (Figure 7.6). The optimal replacement time is 113.48 hours, resulting in the 2.77 cost per unit time. This means that the expected cost is $314.34.

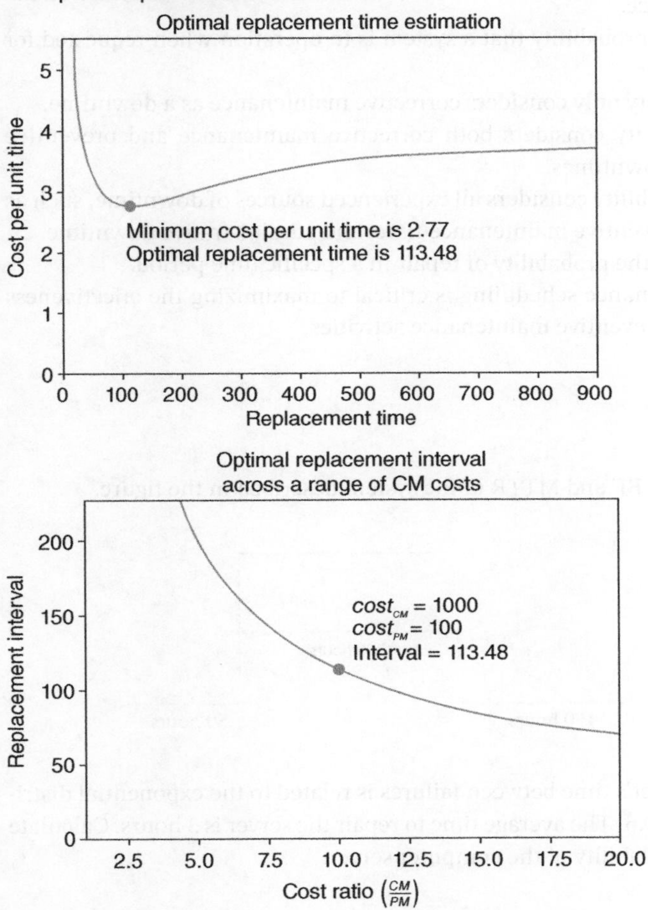

Figure 7.6 The optical replacement interval showing the minimal total cost with Python.

7.8 Summary

- Corrective maintenance focuses on correcting the failures of the components.
- Preventive maintenance is an activity to replace or repair components before they fail.

- The mean time between failures (MTBF) is a measure to quantify the amount of operation times between failures.
- The mean time to repair (MTTR) is a measure to evaluate the efficiency of corrective maintenance.
- Availability is the probability that a system is in operation when requested for use.
- Inherent availability only considers corrective maintenance as a downtime.
- Achieved availability considers both corrective maintenance and preventive maintenance as downtimes.
- Operational availability considers all experienced sources of downtime, such as corrective and preventive maintenance time, and administrative downtime.
- Maintainability is the probability of repair in a specific time period.
- Preventive maintenance scheduling is critical to maximizing the effectiveness and economy of preventive maintenance activities.

Exercises

1 Calculate the MTBF and MTTR of the system described in the figure.

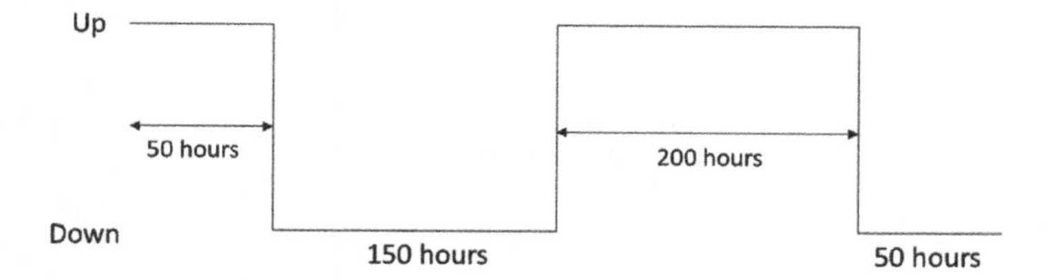

2 A computer server's time between failures is related to the exponential distribution with $\lambda = 0.6$. The average time to repair the server is 3 hours. Calculate the inherent availability of the computer server.

3 A system exhibits a constant failure rate of 1 failure per 500 hours of operation. The average corrective maintenance duration is 10 hours. Preventive maintenance is conducted every 200 hours of operation. The average preventive maintenance duration is 5 hours. Calculate the achieved availability of a system.

4 A system exhibits a constant failure rate of 1 failure per 500 hours of operation. The average corrective maintenance duration is 10 hours. Preventive maintenance is conducted every 200 hours of operation. The average preventive maintenance duration is 5 hours. Calculate the achieved availability of the system. The logistics and administrative delay time is 10 hours. Calculate the operational availability of the system.

5 A system consists of 3 components in parallel. Each component has the MTBF as 500 hours and MTTR as 5 hours. Calculate the system's inherent availability.

6 A system's repair time is normally distributed with a mean of 10 hours and a standard deviation of 3 hours. The average crew size necessary to repair the system is 3 members. The failure occurred every 150 hours. The design goal is that the MH/OH does not exceed 3 corrective maintenance hours for every 100 operating hours. Calculate the MH/OH and determine whether the design goal is met.

7 A system's repair time is normally distributed with a mean of 10 hours and a standard deviation of 3 hours. The average crew size necessary to repair the system is 3 members. Failure occurred every 150 hours. Preventive maintenance is conducted every 100 operating hours. The mean preventive maintenance time is 2 hours. The design goal was that the MH/OH does not exceed 3 corrective maintenance hours for every 100 operating hours. Calculate the MH/OH and determine whether the design goal was met.

8 A system's failure characteristics are modeled by the normal distribution with $\mu = 100$ hours and $\sigma = 30$ hours. If failure occurs, the cost of repairing and downtime of the system is $5,000. For preventive maintenance, the cost of replacement is $500. The system operates 5000 hours per year. The reliability engineer wants to determine whether preventive maintenance by every 300 hours or 600 hours is the more economic schedule. Calculate the total cost, and determine the schedule.

8

Case Studies

Chapter Overview and Learning Objectives

- To understand the concept of parametric and nonparametric reliability analyses.
- To apply Minitab and Python to conduct parametric and nonparametric reliability analyses.
- To understand the objectives of warranty analysis.
- To apply Minitab to conduct a warranty analysis.
- To understand stress–strength interference and apply both Minitab and Python to construct the plots and compute the probability of stress failure.

8.1 Parametric Reliability Analysis

Parametric reliability analysis assumes that the reliability data follows certain types of statistical distributions. For example, a product's failure data could be well described by a lognormal distribution. In that case, it is part of parametric reliability analysis or parametric distribution analysis.

In previous chapters, we learned that there were different types of failure times:

- Exact failure time
- Right-censored data
- Interval-censored data
- Left-censored data

More detailed descriptions are given in previous chapters. A point here is that depending on failure types, computation will be varied, and we also need to select functions or write codes in Minitab and Python properly. Based on the next case

Reliability Analysis Using MINITAB and Python, First Edition. Jaejin Hwang.
© 2023 John Wiley & Sons, Inc. Published 2023 by John Wiley & Sons, Inc.
Companion Website: www.wiley.com\go\Hwang\ReliabilityAnalysisUsingMinitabandPython

study, we could learn the research questions to be explored and the approaches using Minitab and Python to answer the questions.

8.1.1 Description of Case Study

Table 8.1 shows the time to failure (hour) of tires used on loader machines in mining and the axle on which the tires were installed (Nouri Qarahasanlou, 2016). The failure data was collected based on field observation from 2003 to 2014. Table 8.1 shows 10 sample data entries of the loader tires.

Table 8.1 Time to failure data of loader tires.

Time to Failure (hour)	Axle
12,870	F
6050	F
7700	F
9000	F
9000	F
6543	B
10,736	B
11,893	B
13,103	B
10,816	F

Note: F = front axle; B = back axle.

Based on this failure data of loader tires, we could come up with several research questions.

- Which probability distributions would well represent the failure characteristics of a loader tire?
- Would failure characteristics of a loader tire be different by axle?
- By what hour will 25% of a loader tire fail?

8.1.2 Minitab Practice

Step 1: Open Minitab and enter the data in the Worksheet (Figure 8.1).

Worksheet 1 ***

↓	C1	C2-T
	Time to Failure (hr)	Axle
1	12870	F
2	6050	F
3	7700	F
4	9000	F
5	9000	F
6	6543	B
7	10736	B
8	11893	B
9	13103	B
10	10816	F

Figure 8.1 Data entry in Minitab Worksheet.

Step 2: Select Reliability/Survival > Distribution Analysis (Right Censoring) > Distribution ID Plot (Figure 8.2).

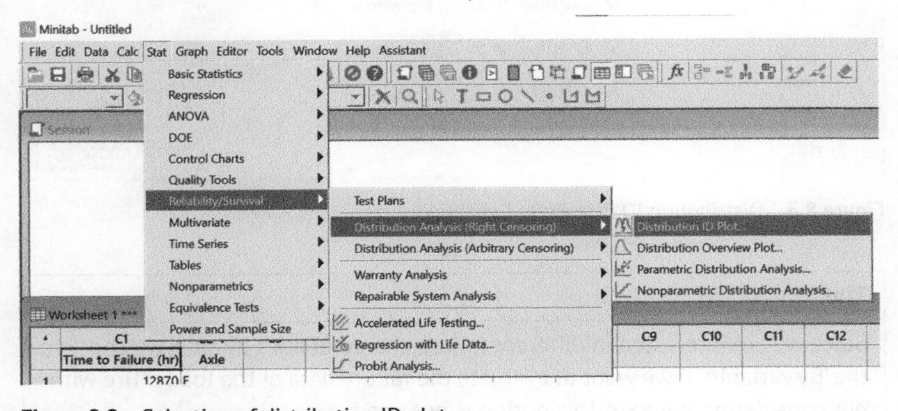

Figure 8.2 Selection of distribution ID plot.

The purpose of this procedure is to identify a proper probability distribution that well describes the failure date of the loader tire. Multiple distributions can be explored and compared to find a proper fit.

Tip!
Our data has the exact failure time. In Minitab, we choose the right-censoring option when we have the exact failure time or right-censored failure data.

Step 3: Select 'Time to Failure (hr)' in the Variables. Check the box of 'By variable.' Select 'Axle.' Select 'Specify.' Four distributions (Weibull, Lognormal, Exponential, Normal) are selected as a default setting. Hit 'OK' (Figure 8.3).

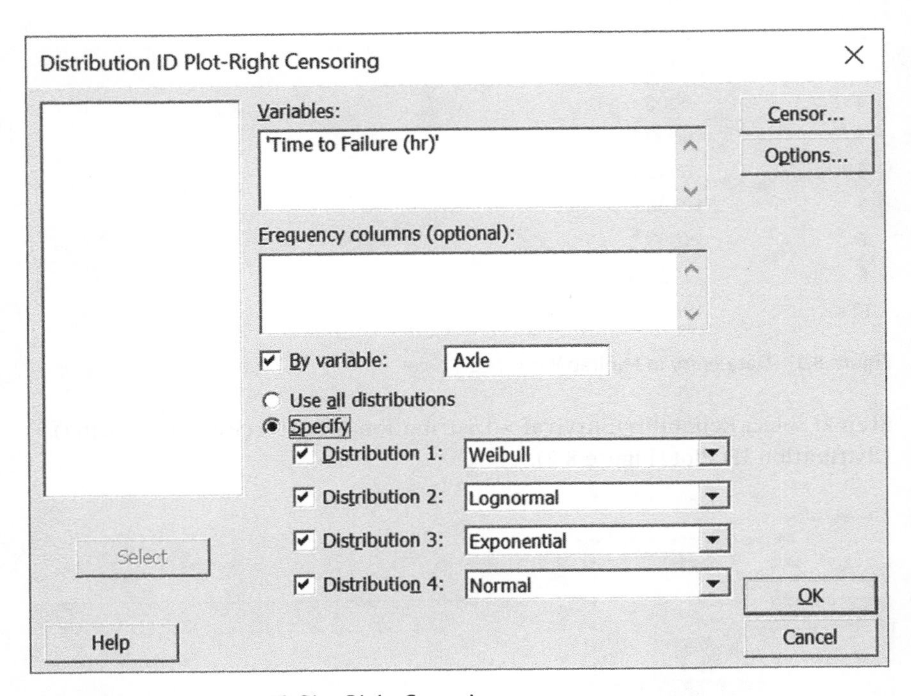

Figure 8.3 Distribution ID Plot-Right Censoring.

Tip!
Since we are interested in different failure characteristics by axle, we checked the 'By variable.' If we want to evaluate the failure data of the loader tire without considering the axle, this option is not needed.

Step 4: You would find a 'Probability Plot for Time to Failure (hr)' by four different distributions (Weibull, Lognormal, Exponential, Normal) (Figure 8.4). The first way to identify the proper fitting is a visual examination. If the data points are well aligned with the straight line, it indicates a good fit for the distributions. Since we tested the probability plot by axle, we could find two

different lines for each distribution. Based on visual examination, we could see data highly deviated from the straight line for the exponential distribution. These deviations could be excluded from consideration. The second way to find a proper fit is to examine the Anderson–Darling (adj) values. The smaller Anderson–Darling (adj) values indicate a better fitness of the distribution to our data. By comparing Anderson–Darling (adj) values among different distributions, we could determine that the Weibull distribution could be the best fit to our data on the loader tire.

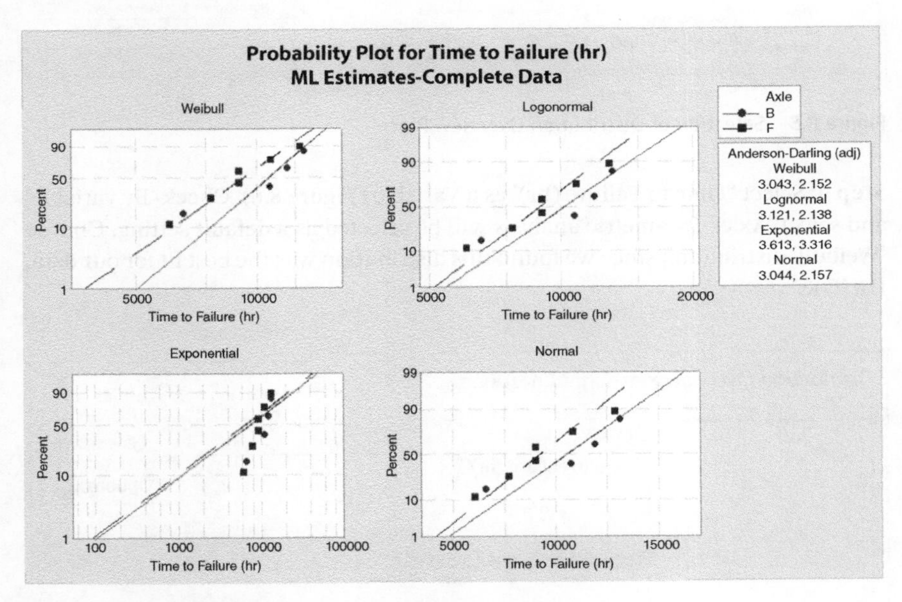

Figure 8.4 Distribution ID plot for time to failure (hour).

Tip!
Depending on the Minitab version, the *p*-value might be shown instead of Anderson–Darling (adj) values. If the *p*-value is typically less than 0.05 or 0.10, this indicates that the distribution does not fit the data well. In other words, a greater *p*-value demonstrates better fitness of the distribution to the data.

Step 5: Select Reliability/Survival > Distribution Analysis (Right Censoring) > Distribution Overview Plot (Figure 8.5). This will allow you to see multiple charts based on the particular distribution you decided to choose.

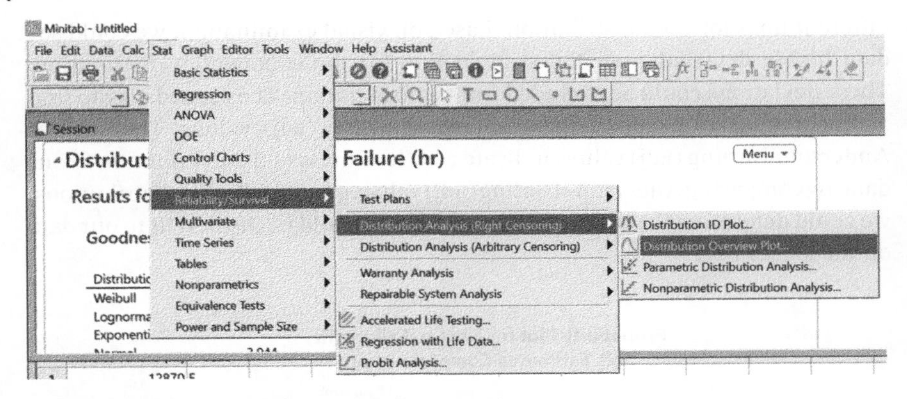

Figure 8.5 Selection of Distribution Overview Plot.

Step 6: Select 'Time to Failure (hr)' as a Variable (Figure 8.6). Check 'By variable' and select 'Axle.' Parametric analysis will be selected as a default setting. Choose 'Weibull' distribution since we found this distribution was the best fit for our data. Hit 'OK.'

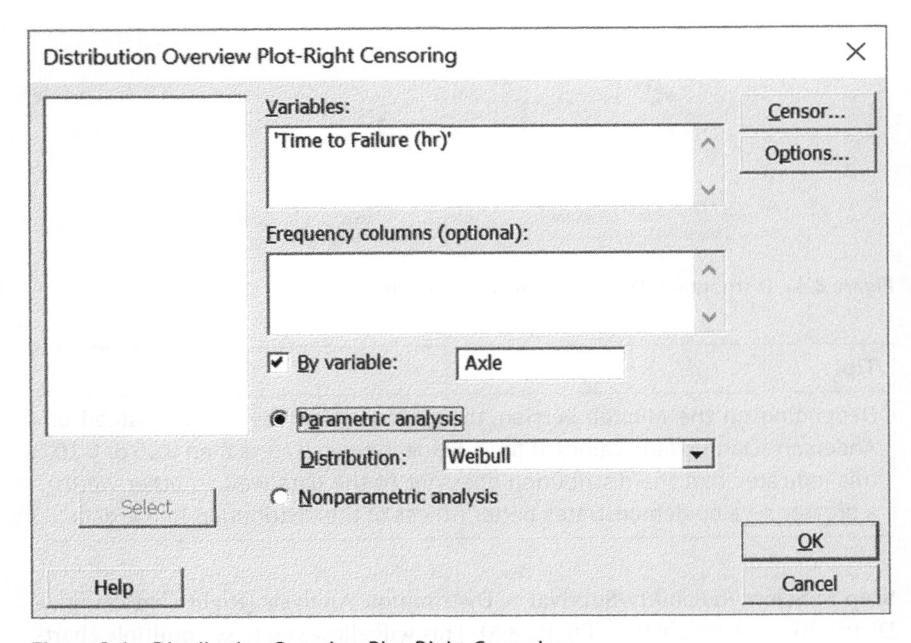

Figure 8.6 Distribution Overview Plot-Right Censoring.

Step 7: You will see four different plots, including PDF plot, probability plot, survival function plot, and hazard function plot (Figure 8.7). By looking at the survival function plot, we can identify that the loader tire installed at the back revealed higher reliability than the tire installed at the front.

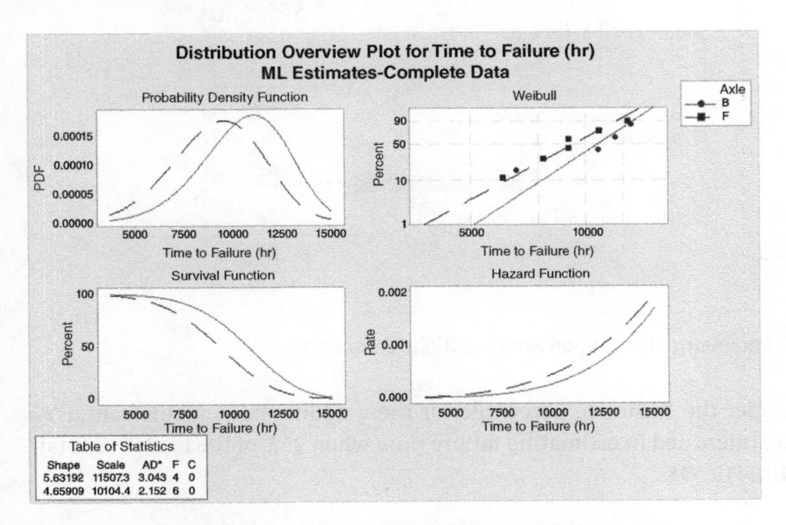

Figure 8.7 Distribution overview plot for the time to failure (hour).

Step 8: If we want to estimate a specific value, we can explore another function. Select Reliability/Survival > Distribution Analysis (Right Censoring) > Parametric Distribution Analysis (Figure 8.8).

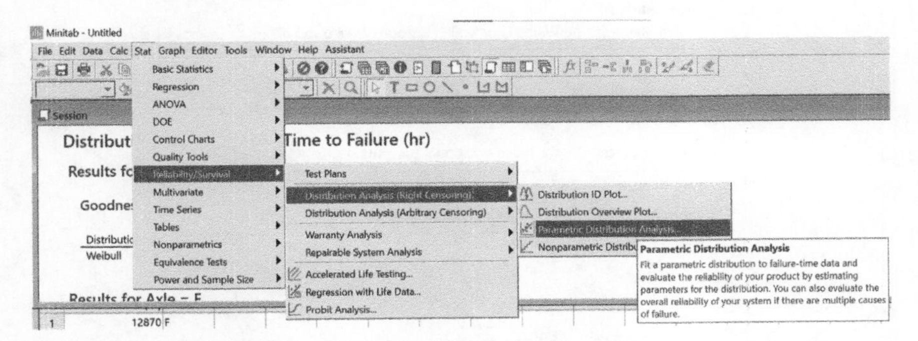

Figure 8.8 Selecting Parametric Distribution Analysis.

Step 9: Select 'Time to Failure (hr)' as a Variable (Figure 8.9). Check 'By variable' and select 'Axle.' Choose 'Weibull' distribution. Since we are interested in estimating a particular time to failure regarding reliability of 80%, select the 'Estimate' tab.

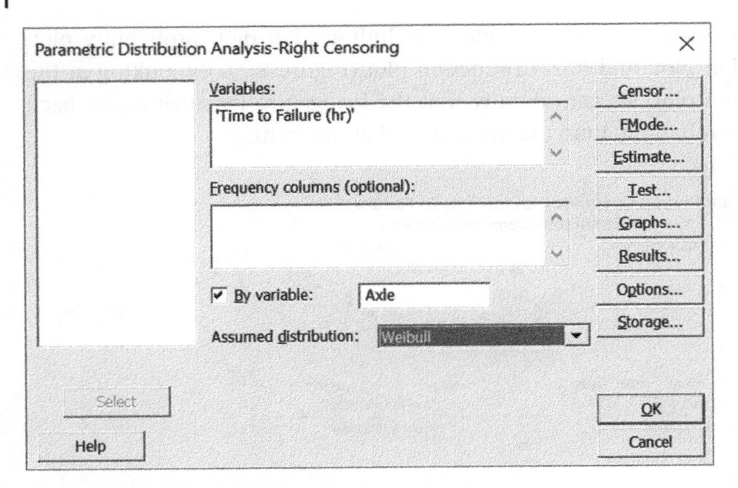

Figure 8.9 Parametric Distribution Analysis-Right Censoring.

Step 10: Under the 'Estimate percentile for these additional percents:' enter 25 since we are interested in estimating failure time when 25% of the loader tires fail (Figure 8.10). Hit 'OK.'

Parametric Distribution Analysis: Estimate

Estimation Method
Maximum Likelihood
Least Squares (failure time(X) on rank(Y))
Assume common shape (slope-Weibull) or scale (1/slope-other dists)
Bayes Analysis
Set shape (slope-Weibull) or scale (1/slope-other dists) at:
Set threshold at:
Estimate percentiles for these additional percents:
25
Estimate probabilities for these times (values):
Estimate survival probabilities
Estimate cumulative failure probabilities
Select
Confidence level: 95.0
Confidence intervals: Two-sided
Help
OK
Cancel

Figure 8.10 Parametric Distribution Analysis: Estimate.

Step 11: You will return to the original window of Parametric Distribution Analysis-Right Censoring. Select the 'Graphs' tab to customize the output graphs that you would like to see to answer our research question (Figure 8.11).

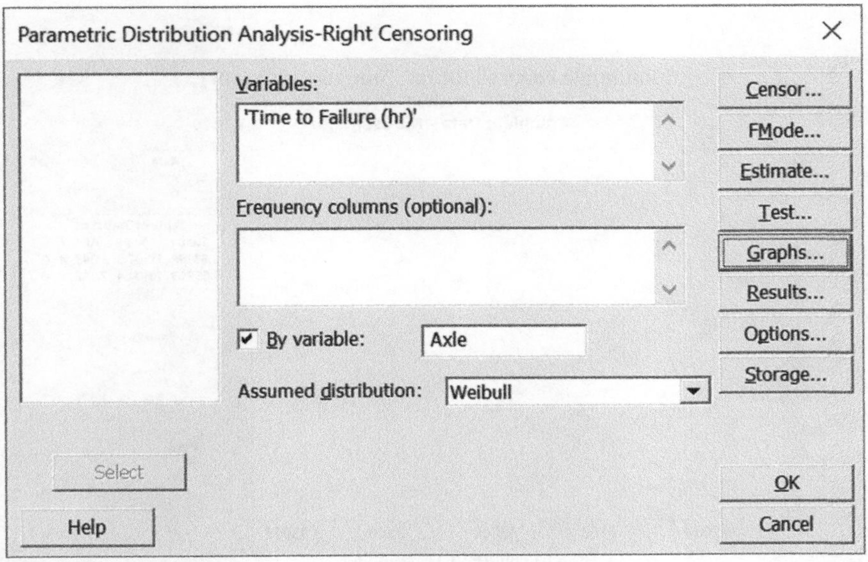

Parametric Distribution Analysis-Right Censoring

Variables: 'Time to Failure (hr)'

Frequency columns (optional):

By variable: Axle

Assumed distribution: Weibull

Censor... FMode... Estimate... Test... Graphs... Results... Options... Storage...

Select Help OK Cancel

Figure 8.11 Parametric Distribution Analysis-Right Censoring.

Step 12: We can only select 'Cumulative failure plot' since we are interested in the failure time regarding cumulative 20% failure of the loader tire (Figure 8.12). Hit 'OK.'

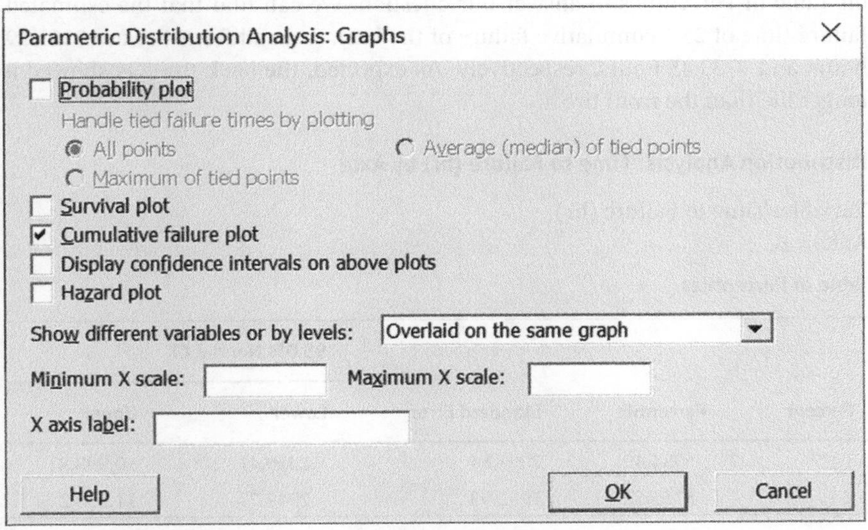

Parametric Distribution Analysis: Graphs

Probability plot

Handle tied failure times by plotting

All points

Average (median) of tied points

Maximum of tied points

Survival plot

Cumulative failure plot

Display confidence intervals on above plots

Hazard plot

Show different variables or by levels: Overlaid on the same graph

Minimum X scale: Maximum X scale:

X axis label:

Help OK Cancel

Figure 8.12 Parametric Distribution Analysis: Graphs.

Step 13: After hitting 'OK,' you could encounter the Cumulative Failure Plot for Time to Failure (hr) (Figure 8.13). We could identify that the loader tire installed at the front continuously shows a higher failure probability than the tire installed at the back.

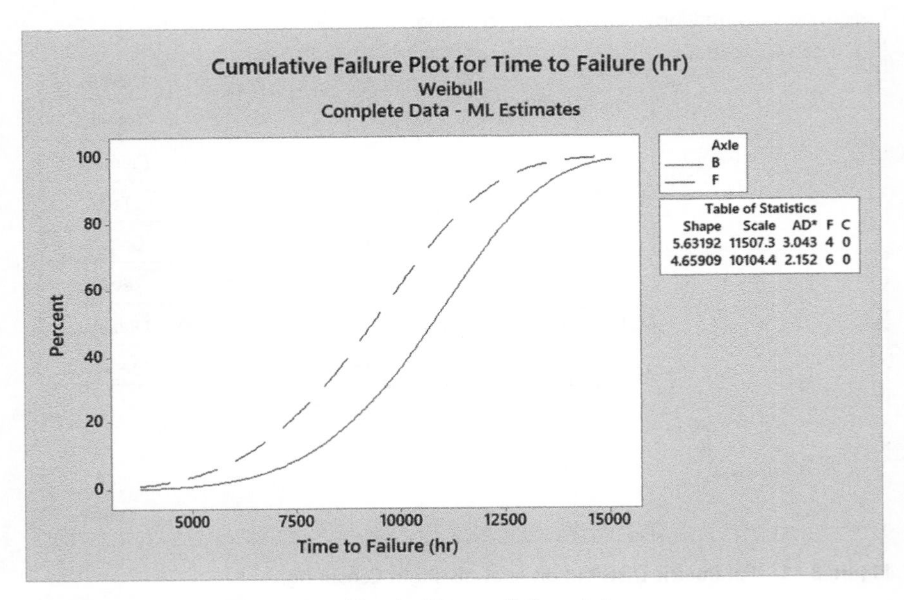

Figure 8.13 Cumulative Failure Plot for Time to Failure (hr).

Step 14: For the session output, we can find the results by the axle. We can focus on Table of Percentiles to answer our question. We can find that the estimated failure time of 25% cumulative failure of the back tire and front tire are 9223.59 hours and 7733.45 hours, respectively. As expected, the back tire has showed a longer life than the front tire.

Distribution Analysis: Time to Failure (hr) by Axle

Variable: Time to Failure (hr)
Axle = B

Table of Percentiles

			95.0% Normal CI	
Percent	**Percentile**	**Standard Error**	**Lower**	**Upper**
1	5084.45	2001.64	2350.41	10,998.8
2	5755.52	1962.93	2949.72	11,230.2

(Continued)

(Continued)

			95.0% Normal CI	
Percent	**Percentile**	**Standard Error**	**Lower**	**Upper**
3	6190.78	1921.40	3369.47	11,374.4
4	6521.18	1882.27	3703.71	11,482.0
5	6791.01	1845.89	3986.26	11,569.2
6	7020.98	1812.01	4233.65	11,643.4
7	7222.58	1780.31	4455.32	11,708.6
8	7402.89	1750.48	4657.22	11,767.3
9	7566.57	1722.29	4843.39	11,820.8
10	7716.89	1695.54	5016.70	11,870.4
20	8816.78	1481.64	6342.60	12,256.1
25	**9223.59**	**1398.75**	**6851.97**	**12,416.1**
30	9582.46	1326.74	7305.05	12,569.9
40	10213.5	1209.29	8098.29	12,881.3
50	10782.3	1123.95	8789.89	13,226.4
60	11330.1	1072.90	9410.86	13,640.8
70	11892.9	1065.03	9978.45	14,174.8
80	12522.0	1119.92	10,508.6	14,921.1
90	13344.1	1292.05	11,037.5	16,132.7
91	13450.6	1321.71	11,094.2	16,307.3
92	13565.1	1355.30	11,152.7	16,499.4
93	13689.7	1393.74	11,213.4	16,713.0
94	13827.4	1438.35	11,277.1	16,954.5
95	13982.4	1491.16	11,345.0	17,232.9
96	14161.9	1555.49	11,419.1	17,563.7
97	14378.8	1637.44	11,502.4	17,974.5
98	14660.9	1750.32	11,602.2	18,526.1
99	15091.8	1934.87	11,738.5	19,403.1

Distribution Analysis: Time to Failure (hr) by Axle

Variable: Time to Failure (hr)

Axle = F

Table of Percentiles

			95.0% Normal CI	
Percent	**Percentile**	**Standard Error**	**Lower**	**Upper**
1	3764.51	1322.96	1890.48	7496.27
2	4373.13	1338.62	2400.14	7967.95
3	4775.99	1335.90	2760.40	8263.32
4	5085.80	1327.65	3048.97	8483.33
5	5341.26	1317.23	3294.01	8660.90
6	5560.68	1305.90	3509.34	8811.08
7	5754.26	1294.23	3702.88	8942.09
8	5928.35	1282.49	3879.65	9058.90
9	6087.16	1270.83	4043.04	9164.78
10	6233.65	1259.33	4195.47	9261.98
20	7323.07	1157.04	5372.84	9981.19
25	**7733.45**	**1114.16**	**5830.96**	**10,256.7**
30	8098.63	1076.05	6241.87	10,507.7
40	8747.71	1012.83	6971.73	10,976.1
50	9339.96	966.857	7624.83	11,440.9
60	9916.54	940.604	8234.22	11,942.6
70	10515.1	940.449	8824.35	12,529.7
80	11191.0	980.771	9424.75	13,288.2
90	12085.2	1103.76	10,104.4	14,454.2
91	12201.8	1125.34	10,184.0	14,619.4
92	12327.5	1149.93	10,267.7	14,800.5
93	12464.6	1178.25	10,356.6	15,001.7
94	12616.2	1211.37	10,452.0	15,228.6
95	12787.4	1250.89	10,556.4	15,489.9
96	12986.1	1299.49	10,673.4	15,800.0
97	13226.9	1362.07	10,809.5	16,185.0
98	13541.3	1449.36	10,978.7	16,701.9
99	14023.8	1594.51	11,222.3	17,524.5

8.1.3 Python Practice

Use the module 'reliability.Fitters' to apply different probability distributions. Since we are interested in four different distributions, we can write:

- Fit_Weibull_2P
- Fit_Lognormal_2P
- Fit_Exponential_1P
- Fit_Normal_2P

We can also import 'matplotlib.pyplot.' This brings a collection of functions like MATLAB, and it helps to create a variety of figures. Then, we can simply call this 'plt.'

Tip!

Not only are there the four distributions mentioned here, but there are also more distributions to be explored. Here is a list of functions for distributions. P indicates the number of parameters. For example, Fit_Exponential_2P denotes the exponential distribution with two parameters.

- Fit_Exponential_1P
- Fit_Exponential_2P
- Fit_Weibull_2P
- Fit_Weibull_3P
- Fit_Gamma_2P
- Fit_Gamma_3P
- Fit_Lognormal_2P
- Fit_Lognormal_3P
- Fit_Loglogistic_2P
- Fit_Loglogistic_3P
- Fit_Normal_2P
- Fit_Gumbel_2P
- Fit_Beta_2P

Enter the data of the loader tires. Since we want to evaluate the failure characteristics of the loader tires by axle, we can define each axle data point, respectively. The 'data_F' means time to failure (hr) data of the loader tire installed at the front. The 'data_B' indicates the failure time of the loader tire installed at the back.

Since we want to see four different probability plots altogether, we can use the 'plt.subplot' function. This allows us to show multiple plots in the same figure. We would find three numbers right after plt.subplot. The first number denotes the number of rows, and the second number denotes the number of columns, and the third number of denotes the index position on a grid. Here, we want to build a 2×2 grid, and the first plot will be located in the upper left corner.

We can start examining the distribution fit using the Weibull distribution. Since we try two different data sets by axle, we could write two separate lines. The 'failures =' code indicates the data set we want to explore. The 'label' code helps us to distinguish two different data sets in the same plot.

For the remaining three distributions, we can apply a similar approach as seen in the figure. The position of plt.subplot is changed for each plot.

It is time to display all four plots together. The 'tight_layout()' function would automatically adjust the subplot size to ideally locate four plots in a single figure. The 'show()' function would eventually display the figure. This is a snippet of the whole code, which is shown in Figure 8.14.

After running all the codes, you would expect to see the figure showing probability plots using four different distributions (Figure 8.15). The result looks similar to the plots created from Minitab.

We could access the output displaying the parameter estimates and goodness of fit results (Figure 8.16). Similar to the Minitab approach, we could examine the

```
from reliability.Fitters import Fit_Weibull_2P, Fit_Lognormal_2P, Fit_Exponential_1P, Fit_Normal_2P
import matplotlib.pyplot as plt

data_F = [12870,6050,7700,9000,9000,10816]
data_B = [6543,10736,11893,13103]

plt.subplot(221)
wb = Fit_Weibull_2P(failures=data_F,label='F')
wb = Fit_Weibull_2P(failures=data_B,label='B')

plt.subplot(222)
lg = Fit_Lognormal_2P(failures=data_F,label='F')
lg = Fit_Lognormal_2P(failures=data_B,label='B')

plt.subplot(223)
ex = Fit_Exponential_1P(failures=data_F,label='F')
ex = Fit_Exponential_1P(failures=data_B,label='B')

plt.subplot(224)
nm = Fit_Normal_2P(failures=data_F,label='F')
nm = Fit_Normal_2P(failures=data_B,label='B')

plt.tight_layout()
plt.show()
```

Figure 8.14 Complete Python code for distribution ID plot.

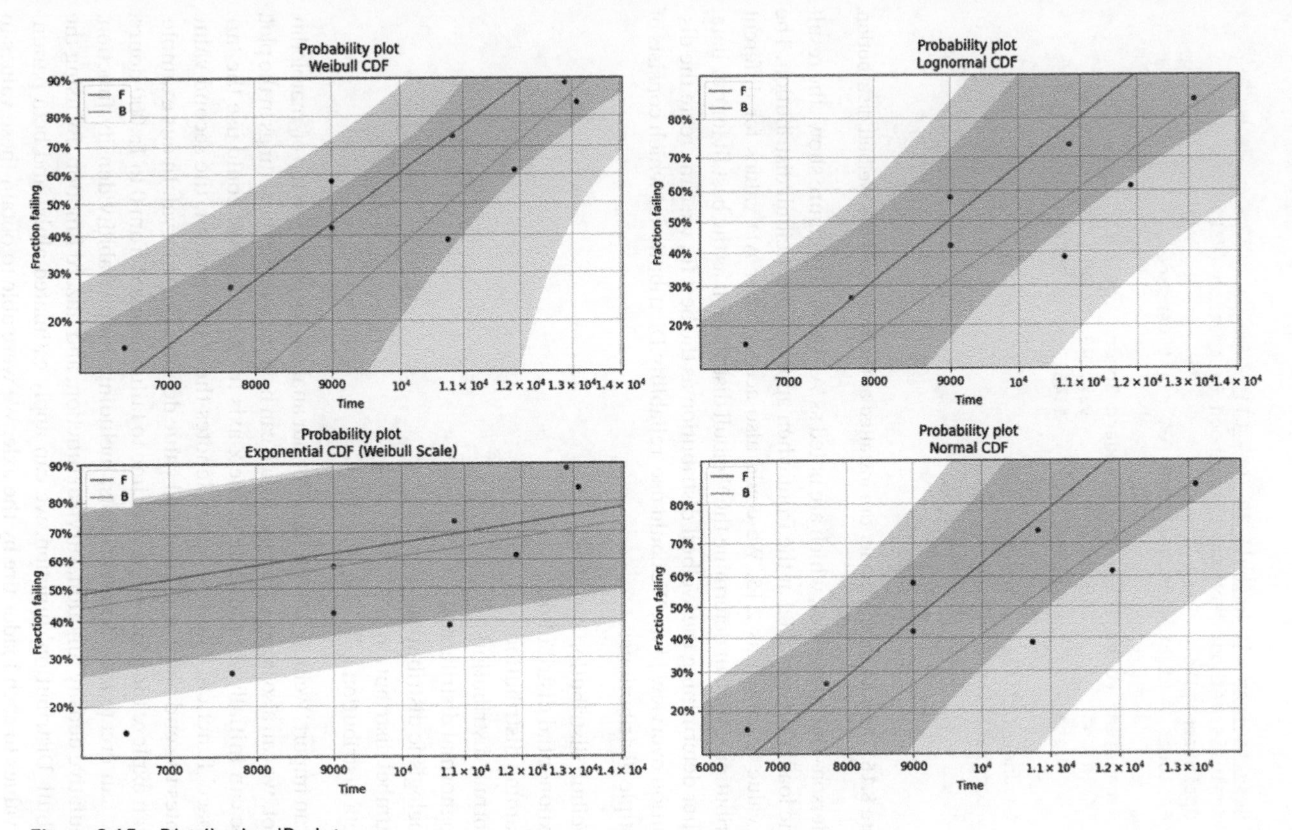

Figure 8.15 Distribution ID plot.

```
Results from Fit_Weibull_2P (95% CI):
Analysis method: Maximum Likelihood Estimation (MLE)
Optimizer: TNC
Failures / Right censored: 6/0 (0% right censored)

Parameter  Point Estimate  Standard Error  Lower CI  Upper CI
    Alpha         10104.4         937.249    8424.7   12118.9
     Beta         4.65908         1.46236   2.51844   8.61925

Goodness of fit    Value
 Log-likelihood -54.6753
           AICc  117.351
            BIC  112.934
             AD  2.15207
```

Figure 8.16 Results for the loader tire installed at the front with the Weibull distribution.

Anderson–Darling values, which are noted as 'AD' here. The figure shows the result of the loader tire installed at the front when applying the Weibull distribution. The AD value is shown as 2.15. We could also access other AD values for different distributions. We can determine the Weibull distribution as the best fit for our data.

After determining the Weibull distribution as the best fit, we can create the distribution overview plots. We could use 'reliability.Distributions,' which consists of multiple distributions:

- Weibull distribution
- Exponential distribution
- Gamma distribution
- Normal distribution
- Lognormal distribution
- Loglogistic distribution
- Gumbel distribution
- Beta distribution

We can import 'Weibull_Distribution' for our analysis purposes. On top of 'matplotlib.pyplot,' we can also import 'numpy,' which can help us to adjust the settings in the plot.

We can initially set the scale of the x-axis in the chart. We could use the 'np.linspace' function. The first value denotes the starting point, the second value denotes the end point, and the third value denotes the interval. In our example, we can explore the plots using the time to failure ranged from 0 to 15,000 hours.

We can create four different plots, including the probability density function, cumulative density function, survival function, and hazard function. By using the 'Weibull_Distribution' function, we can apply estimated alpha and beta parameter values to each loader tire by the axle. We were able to obtain these values in previous outputs when analyzing the Distributing ID plot. The first example shows how we create the probability density function of the front and back loader tires. 'PDF' displays the probability density function.

Similar to the previous approach, we can create three plots, including the cumulative density function, survival function, and hazard function. The 'CDF,' 'SF,' and 'HF' are related functions.

Last, add 'plt.tight_layout()' and 'plt.show()' to display multiple plots in the same figure. Figure 8.17 shows all of the codes for the distribution overview plot. Run the module to see the plots (Figure 8.17).

```
from reliability.Distributions import Weibull_Distribution
import matplotlib.pyplot as plt
import numpy as np

xvals = np.linspace(0,15000,2500)

plt.subplot(221)
dist_F = Weibull_Distribution(alpha=10148.2, beta=4.6807).PDF(xvals=xvals,label='F')
dist_B = Weibull_Distribution(alpha=11507.3, beta=5.63191).PDF(xvals=xvals,label='B')
plt.legend()
plt.title('Probability Density Function')

plt.subplot(222)
dist_F = Weibull_Distribution(alpha=10148.2, beta=4.6807).CDF(xvals=xvals,label='F')
dist_B = Weibull_Distribution(alpha=11507.3, beta=5.63191).CDF(xvals=xvals,label='B')
plt.legend()
plt.title('Cumulative Density Function')

plt.subplot(223)
dist_F = Weibull_Distribution(alpha=10148.2, beta=4.6807).SF(xvals=xvals,label='F')
dist_B = Weibull_Distribution(alpha=11507.3, beta=5.63191).SF(xvals=xvals,label='B')
plt.legend()
plt.title('Survival Function')

plt.subplot(224)
dist_F = Weibull_Distribution(alpha=10148.2, beta=4.6807).HF(xvals=xvals,label='F')
dist_B = Weibull_Distribution(alpha=11507.3, beta=5.63191).HF(xvals=xvals,label='B')
plt.legend()
plt.title('Hazard Function')
```

Figure 8.17 Python codes of the distribution overview plot.

You may find the distribution overview plot (Figure 8.18). These four plots look similar to the plots that we obtained from Minitab.

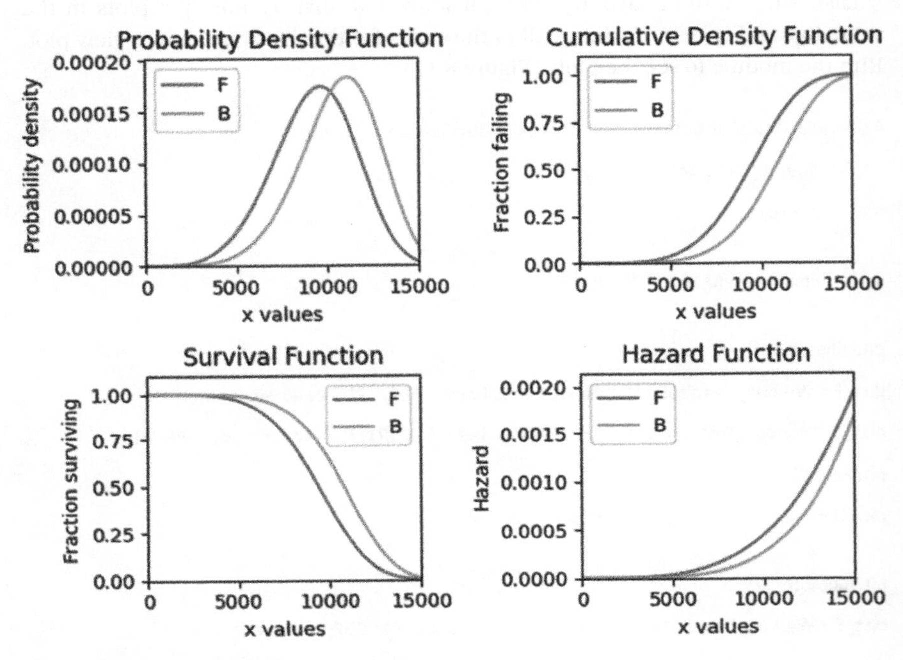

Figure 8.18 The distribution overview plot.

Last, we want to find the point of time that 25% of the loader tires will fail.

We can use 'reliability.Distributions' and import 'Weibull_Distribution.' We could apply estimated parameter values including alpha and beta to each front and back tire, respectively.

We can apply 'inverse_SF' function to answer our question. The SF denotes a survival function. Since we are interested in the cumulative failure of 25%, this could also be written as 75% of survival.

Since we compute the time to failure of each tire, it is time to display the results using the 'print' function. Then, Run Module.

Figure 8.19 lists all the codes for the distribution overview plot.

```
pip install reliability
```

```
[ ] pip install matplotlib==3.1.3
```

```
[ ] from reliability.Distributions import Weibull_Distribution

    dist_F = Weibull_Distribution(alpha=10148.2, beta=4.6807)
    dist_B = Weibull_Distribution(alpha=11507.3, beta=5.63191)
```

```
[ ] tf_F = dist_F.inverse_SF(0.75)
    tf_B = dist_B.inverse_SF(0.75)
```

```
print('Time to failure of 20% cumulative failure of front tire is', tf_F)
print('Time to failure of 20% cumulative failure of back tire is', tf_B)
```

Figure 8.19 Python codes for parameter estimation.

You would find the time to failure of the front and back tires for cumulative 20% failure (Figure 8.20).

Time to failure of 20% cumulative failure of front tire is 7776.59326485261

Time to failure of 20% cumulative failure of back tire is 9223.551528820697

Figure 8.20 The output of time to failure regarding 20% cumulative failure of front and back tires.

Based on the Minitab and Python results, we are able to answer our research questions.

- Which probability distributions would well represent the failure characteristics of the loader tires?
  - The Weibull distribution showed the best fit representing the failure characteristics of the loader tires.
- Would the failure characteristics of the loader tires be different by axle?
  - Yes, the failure characteristics of the loader tires were different by axle. The loader tire installed at the back continuously showed higher reliability than the loader tire installed at the front.

- By what time would 25% of the loader tires fail?
  - For the front tire, 25% of the tires will fail by 7733 to 7777 hours. For the back tire, 25% of the tires will fail by 9224 hours.

8.2 Nonparametric Reliability Analysis

If none of the existing probability distributions fit the life data, nonparametric reliability analysis could be an alternative option. When conducting a nonparametric analysis, there is no assumption of certain probability distribution. The Kaplan–Meier estimator is one of the commonly used methods in the field of nonparametric reliability analysis. We will apply Minitab and Python to conduct our analysis.

8.2.1 Description of Case Study

A reliability engineer at a furnace manufacturer collected the time-to-failure data of furnace components. the exact failure time was obtained for 10 components. Five of the components were still operating at the end of the observation period. Table 8.2 shows the summary of the obtained exact and right-censored failure data.

Table 8.2 Failure data of furnace components.

Failure Time (hours)	Censored
530	No
750	No
1700	No
3870	No
4500	No
4940	No
7230	No
1320	No
1501	No
1600	No
400	Yes
500	Yes
485	Yes
6040	Yes
4670	Yes

8.2.2 Minitab Practice

Figure 8.21 shows the data format when using Minitab.

↓	C1	C2-T
	Failure Time (hours)	Censored
1	530	No
2	750	No
3	1700	No
4	3870	No
5	4500	No
6	4940	No
7	7230	No
8	1320	No
9	1501	No
10	1600	No
11	400	Yes
12	500	Yes
13	485	Yes
14	6040	Yes
15	4670	Yes

Figure 8.21 The exact and right-censored data.

Go to Reliability/Survival > Distribution Analysis (Right Censoring) > Nonparametric Distribution Analysis (Figure 8.22).

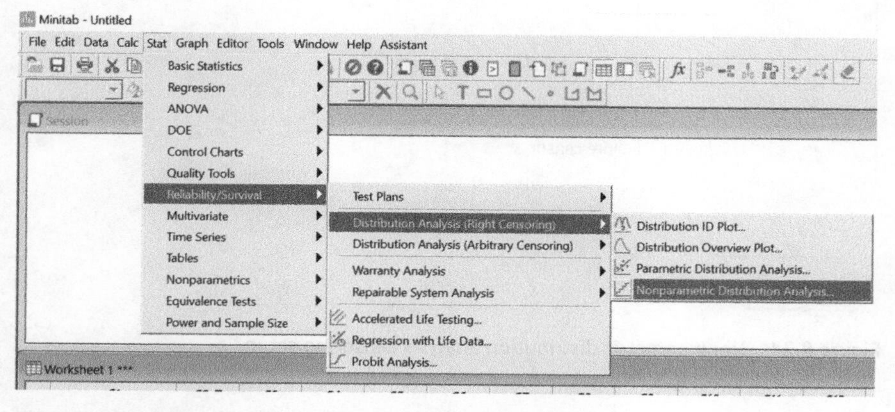

Figure 8.22 Nonparametric distribution analysis option.

Set 'Failure Time (hours)' as Variables. Click 'Censor' (Figure 8.23).

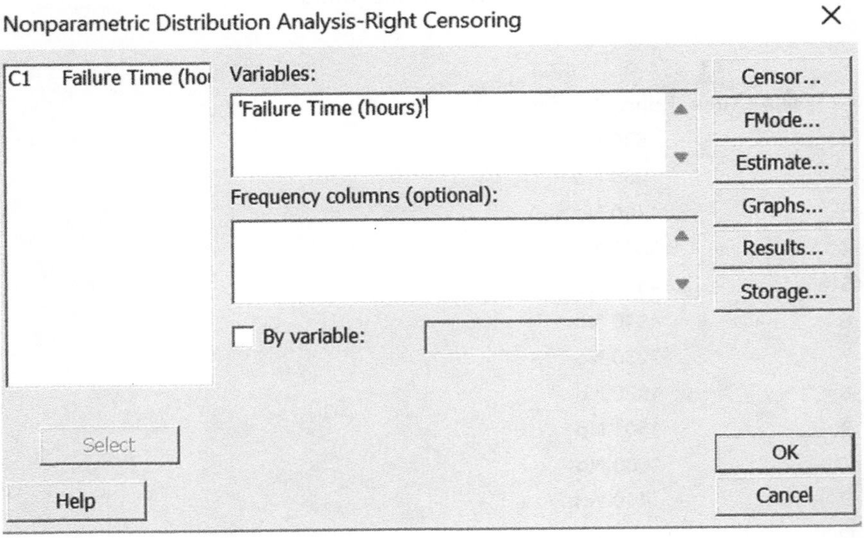

Nonparametric Distribution Analysis-Right Censoring

C1 Failure Time (hoı
Variables:
'Failure Time (hours)'
Frequency columns (optional):
By variable:
Select
Help
Censor...
FMode...
Estimate...
Graphs...
Results...
Storage...
OK
Cancel

Figure 8.23 Nonparametric distribution analysis setup.

Assign 'Censored' column. Type 'Yes' as a censoring value. Click 'OK' (Figure 8.24).

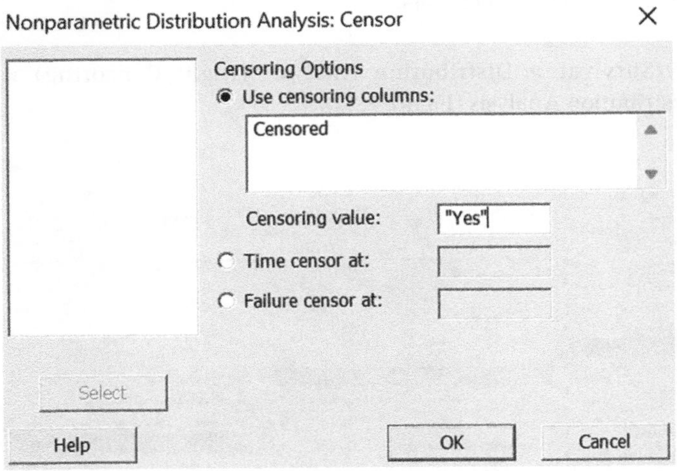

Nonparametric Distribution Analysis: Censor

Censoring Options
Use censoring columns:
Censored
Censoring value: "Yes"
Time censor at:
Failure censor at:
Select
Help
OK
Cancel

Figure 8.24 Nonparametric distribution analysis censoring setup.

Click 'Graph.' Select 'Survival plot.' Click 'OK' (Figure 8.25).

Nonparametric Distribution Analysis: Graphs

- [x] Survival plot
- [ ] Cumulative failure plot
- [ ] Display confidence intervals on plot
- [ ] Hazard plot

Show different variables or by levels: Overlaid on the same graph

Minimum X scale: 0 Maximum X scale:

X axis label:

Help OK Cancel

Figure 8.25 Nonparametric distribution analysis graph setup.

The survival plot is constructed (Figure 8.26).

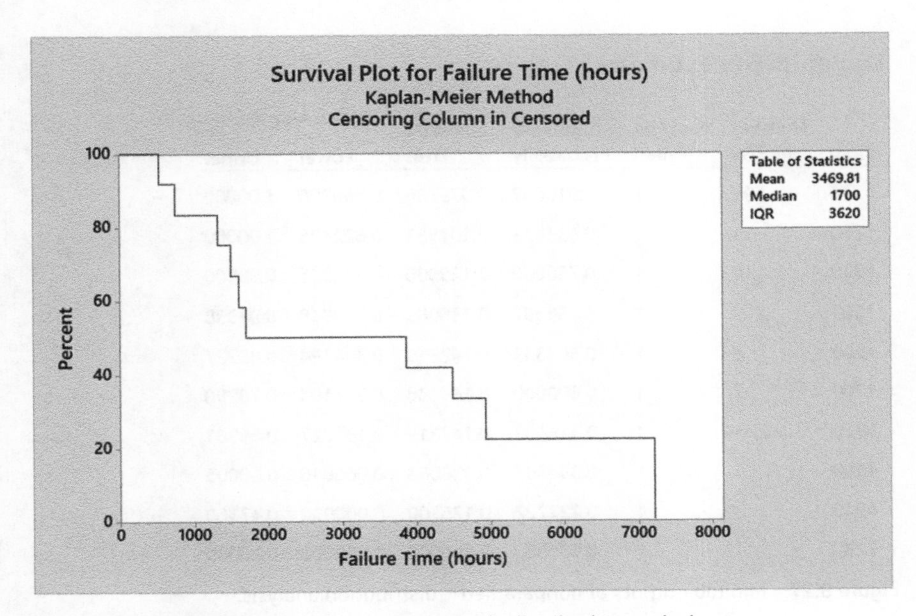

Figure 8.26 Survival plot with a nonparametric distribution analysis.

Here is the output from Minitab (Figure 8.27). It shows that half of the components would survive at 1700 hours.

Variable: Failure Time (hours)

Censoring

Censoring Information	Count
Uncensored value	10
Right censored value	5

Censoring value: Censored = Yes

Nonparametric Estimates

Characteristics of Variable

	Standard	95.0% Normal CI					
Mean(MTTF)	Error	Lower	Upper	Q1	Median	Q3	IQR
3469.81	758.325	1983.52	4956.10	1320	1700	4940	3620

Kaplan-Meier Estimates

	Number	Number	Survival	Standard	95.0% Normal CI	
Time	at Risk	Failed	Probability	Error	Lower	Upper
530	12	1	0.916667	0.079786	0.760290	1.00000
750	11	1	0.833333	0.107583	0.622475	1.00000
1320	10	1	0.750000	0.125000	0.505005	0.99500
1501	9	1	0.666667	0.136083	0.399949	0.93338
1600	8	1	0.583333	0.142319	0.304394	0.86227
1700	7	1	0.500000	0.144338	0.217104	0.78290
3870	6	1	0.416667	0.142319	0.137727	0.69561
4500	5	1	0.333333	0.136083	0.066616	0.60005
4940	3	1	0.222222	0.128300	0.000000	0.47369
7230	1	1	0.000000	0.000000	0.000000	0.00000

Figure 8.27 Minitab outputs of nonparametric distribution analysis.

8.2.3 Python Practice

Python can be used to conduct a nonparametric reliability analysis (Figure 8.28).

'KaplanMeier' function can be imported.

'f' denotes the exact time to failure.

'rc' denotes the right-censored failure time.

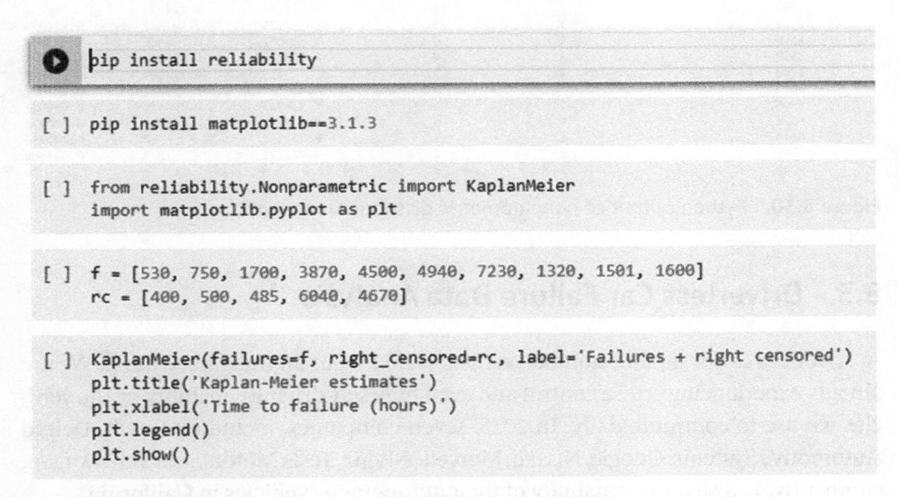

```
pip install reliability
```

```
pip install matplotlib==3.1.3
```

```
from reliability.Nonparametric import KaplanMeier
import matplotlib.pyplot as plt
```

```
f = [530, 750, 1700, 3870, 4500, 4940, 7230, 1320, 1501, 1600]
rc = [400, 500, 485, 6040, 4670]
```

```
KaplanMeier(failures=f, right_censored=rc, label='Failures + right censored')
plt.title('Kaplan-Meier estimates')
plt.xlabel('Time to failure (hours)')
plt.legend()
plt.show()
```

Figure 8.28 Python codes to conduct a nonparametric distribution analysis.

After running all codes, the survival function plot can be constructed (Figure 8.29).

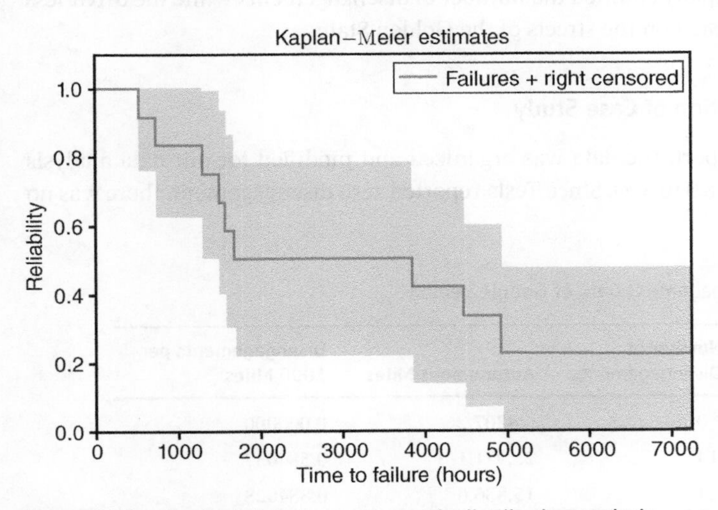

Figure 8.29 Survival plot of nonparametric distribution analysis.

The result is also generated with Python (Figure 8.30). The result was found to be consistent with the Minitab results.

```
Results from KaplanMeier (95% CI):
 Failure times  Censoring code (censored=0)  Items remaining  Kaplan-Meier Estimate  Lower CI bound  Upper CI bound
           400                            0               15                      1               1               1
           485                            0               14                      1               1               1
           500                            0               13                      1               1               1
           530                            1               12               0.916667         0.76029               1
           750                            1               11               0.833333        0.622475               1
          1320                            1               10                   0.75        0.505005        0.994995
          1501                            1                9               0.666667        0.399949        0.933384
          1600                            1                8               0.583333        0.304394        0.862273
          1700                            1                7                    0.5        0.217104        0.782896
          3870                            1                6               0.416667        0.137727        0.695606
          4500                            1                5               0.333333        0.066616        0.600051
          4670                            0                4               0.333333        0.066616        0.600051
          4940                            1                3               0.222222               0        0.473686
          6040                            0                2               0.222222               0        0.473686
          7230                            1                1                      0               0               0
```

Figure 8.30 Python output of nonparametric distribution analysis.

8.3 Driverless Car Failure Data Analysis

A driverless car or an autonomous vehicle is no longer an imaginary thing. We are already experiencing cruise control and even fully self-driving capability by the vehicles we use to commute daily. In 2015, seven companies, including Bosch, Delphi Automotive Systems, Google, Nissan, Mercedes-Benz, Tesla Motors, and Volkswagen, competitively tested the capability of their autonomous vehicles in California.

Human test drivers were inside the cars, and they took control of the cars when they felt it was needed for safety. The human intervention was called a disengagement. Each company counted the number of disengagements while the driverless vehicles were tested on the streets of the Golden State.

8.3.1 Description of Case Study

Based on the report, the data was organized and modified for our data analysis purpose (Tables 8.3 to 8.8). Since Tesla reported zero disengagement, there was no data to analyze.

Table 8.3 Disengagement data of Google vehicle.

Month	Number of Disengagements	Autonomous Miles	Disengagements per 1000 Miles
2014/09	0	4207.2	0.000000
2014/10	14	23,971.1	0.584037
2014/11	14	15,836.6	0.884028

(Continued)

Table 8.3 (Continued)

Month	Number of Disengagements	Autonomous Miles	Disengagements per 1000 Miles
2014/12	40	9413.1	4.249397
2015/01	48	18,192.1	2.638508
2015/02	12	18,745.1	0.640167
2015/03	26	22,204.2	1.170950
2015/04	47	31,927.3	1.472094
2015/05	9	38,016.8	0.236737
2015/06	7	42,046.6	0.166482
2015/07	19	34,805.1	0.545897
2015/08	4	38,219.8	0.104658
2015/09	15	36,326.6	0.412921
2015/10	11	47,143.5	0.233330
2015/11	6	43,275.9	0.138645

Table 8.4 Disengagement data of Nissan vehicle.

Month	Number of Disengagements	Autonomous Miles	Disengagements per 1000 Miles
2014/11	7	112.7	62.111801
2014/12	32	366.6	87.288598
2015/01	31	117.7	263.381478
2015/02	0	16.4	0.000000
2015/03	0	5	0.000000
2015/10	28	374.9	74.686583
2015/11	8	492.1	16.256858

Table 8.5 Disengagement data of Mercedes-Benz vehicle.

Month	Number of Disengagements	Autonomous Miles	Disengagements per 1000 Miles
2014/09	44	107.65	408.732002
2014/10	228	227.03	1004.272563
2014/11	141	365.3	385.984123

(Continued)

Table 8.5 (Continued)

Month	Number of Disengagements	Autonomous Miles	Disengagements per 1000 Miles
2014/12	36	42.81	840.925018
2015/01	48	29.43	1630.988787
2015/02	16	38.9	411.311054
2015/03	39	18.3	2131.147541
2015/04	72	47	1531.914894
2015/05	34	55.1	617.059891
2015/06	127	171.63	739.963876
2015/07	69	118.82	580.710318
2015/08	24	22.17	1082.543978
2015/09	50	19.72	2535.496957
2015/10	17	29.59	574.518418
2015/11	22	43.5	505.747126

Table 8.6 Disengagement data of Volkswagen vehicle.

Month	Number of Disengagements	Autonomous Miles	Disengagements per 1000 Miles
2014/09	18	1026.5	17.535314
2014/10	65	899.75	72.242290
2014/11	66	4088.62	16.142366
2014/12	14	1733.63	8.075541
2015/01	9	749.99	12.000160
2015/02	1	160.94	6.213496
2015/03	0	121.79	0.000000
2015/04	0	3.11	0.000000
2015/05	0	99.42	0.000000
2015/06	1	45.36	22.045855
2015/07	0	31.07	0.000000
2015/08	0	329.95	0.000000
2015/09	1	124.27	8.046994

Table 8.7 Disengagement data of Bosch vehicle.

Month	Number of Disengagements	Autonomous Miles	Disengagements per 1000 Miles
2014/12	126	92.5	1362.162162
2015/01	86	236.2	364.098222
2015/02	21	51.2	410.156250
2015/04	83	131.1	633.104500
2015/05	10	30.8	324.675325
2015/07	40	91.3	438.116101
2015/08	35	108.8	321.691176
2015/09	27	93.1	290.010741
2015/11	94	22.4	4196.428571

Table 8.8 Disengagement data of Delphi vehicle.

Month	Number of Disengagements	Autonomous Miles	Disengagements per 1000 Miles
2014/10	18	397	45.340050
2014/11	20	586	34.129693
2014/12	42	577	72.790295
2015/01	22	242	90.909091
2015/02	2	381	5.249344
2015/03	35	3837	9.121710
2015/04	8	48	166.666667
2015/05	2	32	62.500000
2015/06	6	95	63.157895
2015/07	8	244	32.786885
2015/08	3	566	5.300353
2015/09	5	91	54.945055
2015/10	6	41	146.341463
2015/11	20	20	1000.000000

8.3.2 Minitab Practice

Each company's disengagement per 1000 miles data can be used for fair comparisons. For the zero disengagements in a particular month, we could assign this as a censored value by adding an additional column called Censoring (Figure 8.31).

↓	C1	C2
	DisengagementPer1000Miles	Censoring
1	0.000000	1
2	0.584037	0
3	0.884028	0
4	4.249397	0
5	2.638508	0
6	0.640167	0
7	1.170950	0
8	1.472094	0
9	0.236737	0
10	0.166482	0
11	0.545897	0
12	0.104658	0
13	0.412921	0
14	0.233330	0
15	0.138645	0

Figure 8.31 Disengagement data set of Google vehicle.

The 'Distribution ID Plot-Right Censoring' is conducted to search for the best fitting distribution. 'Censor' option is also set by assigning the column called Censoring (Figure 8.32).

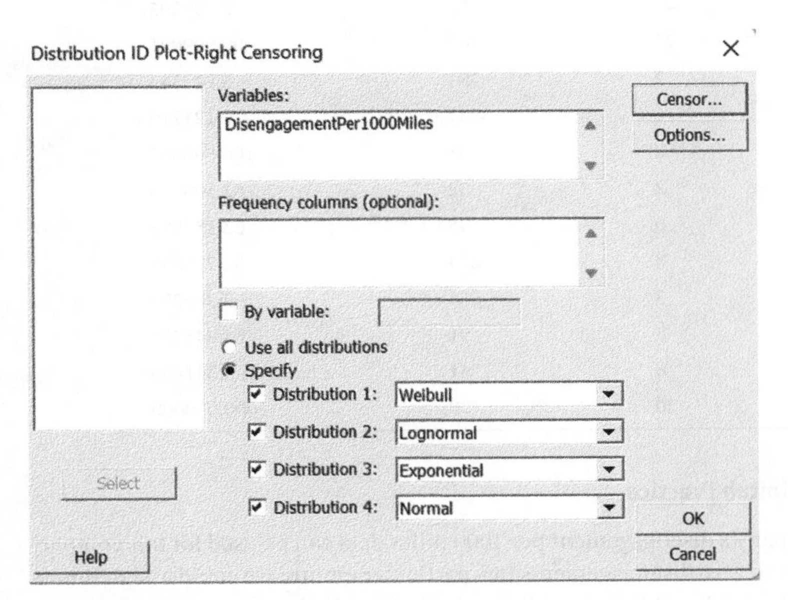

Figure 8.32 Distribution ID plot-right censoring setup for Google data.

Based on the result, the lognormal distribution showed the lowest Anderson–Darling value, and visually examination was also satisfactory (Figure 8.33).

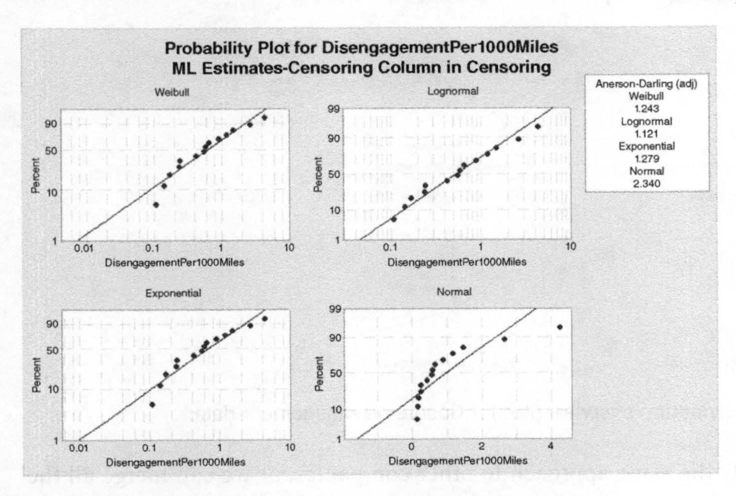

Figure 8.33 Distribution ID plot-right censoring probability plots.

'Distribution Overview Plot-Right Censoring' can be used to provide various reliability charts using the lognormal distribution (Figure 8.34).

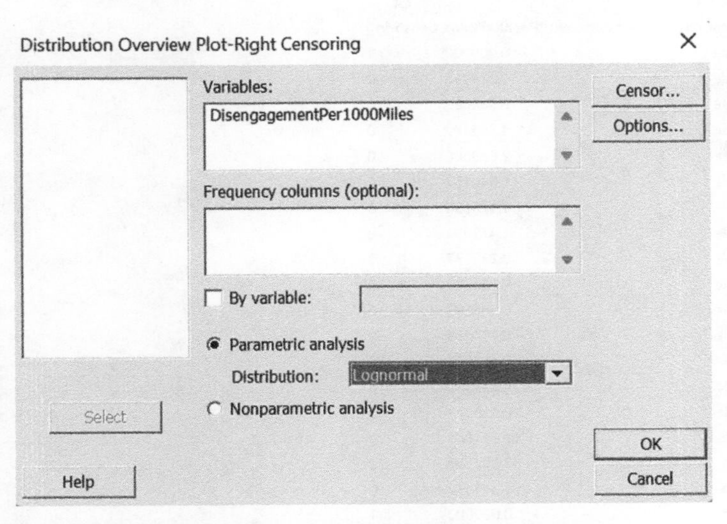

Figure 8.34 Distribution overview plot-right censoring setup.

Four different plots, including PDF, probability plot, $R(t)$, and $h(t)$, are provided (Figure 8.35).

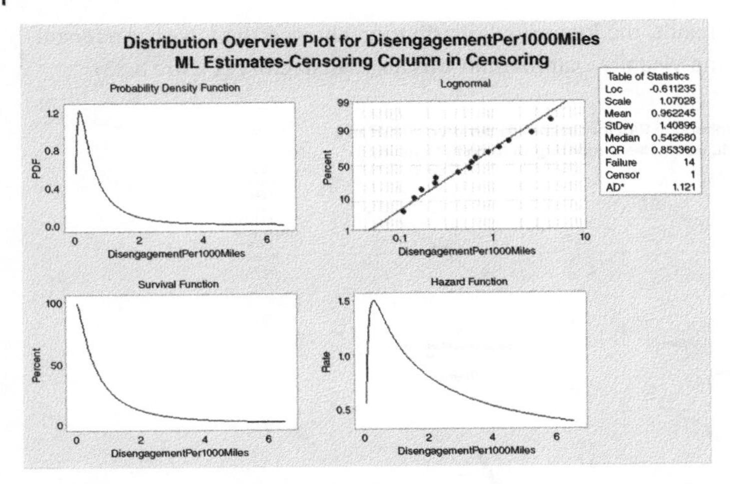

Figure 8.35 Distribution overview plot for Google disengagement data.

We can apply the same approach to other companies, or we can merge all the data into one sheet. Figure 8.36 shows part of the merged data. A Company column is added to indicate each company. The Month column is added to understand each company's cumulative time to test the vehicles.

↓	C1	C2-T	C3	C4
	Month	Company	DisengagementPer1000Miles	Censoring
1	1	Google	0.000000	1
2	2	Google	0.584037	0
3	3	Google	0.884028	0
4	4	Google	4.249397	0
5	5	Google	2.638508	0
6	6	Google	0.640167	0
7	7	Google	1.170950	0
8	8	Google	1.472094	0
9	9	Google	0.236737	0
10	10	Google	0.166482	0
11	11	Google	0.545897	0
12	12	Google	0.104658	0
13	13	Google	0.412921	0
14	14	Google	0.233330	0
15	15	Google	0.138645	0
16	1	Nissan	62.111801	0
17	2	Nissan	87.288598	0
18	3	Nissan	263.381478	0
19	4	Nissan	0.000000	1
20	5	Nissan	0.000000	1
21	6	Nissan	74.686583	0
22	7	Nissan	16.256858	0

Figure 8.36 Merged data set of disengagement data of six companies.

After conducting the 'Distribution ID Plot,' we could see the results of all companies in one chart (Figure 8.37). Based on the visual examination, the lognormal distribution deems the most appropriate fit for all the data.

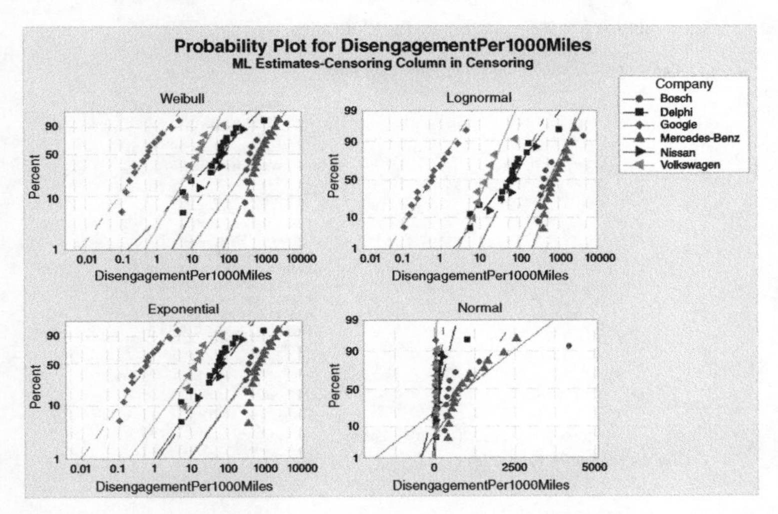

Figure 8.37 Probability plots of four different distributions with six companies.

By applying the lognormal distribution, a distribution overview plot for the six companies can be created (Figure 8.38).

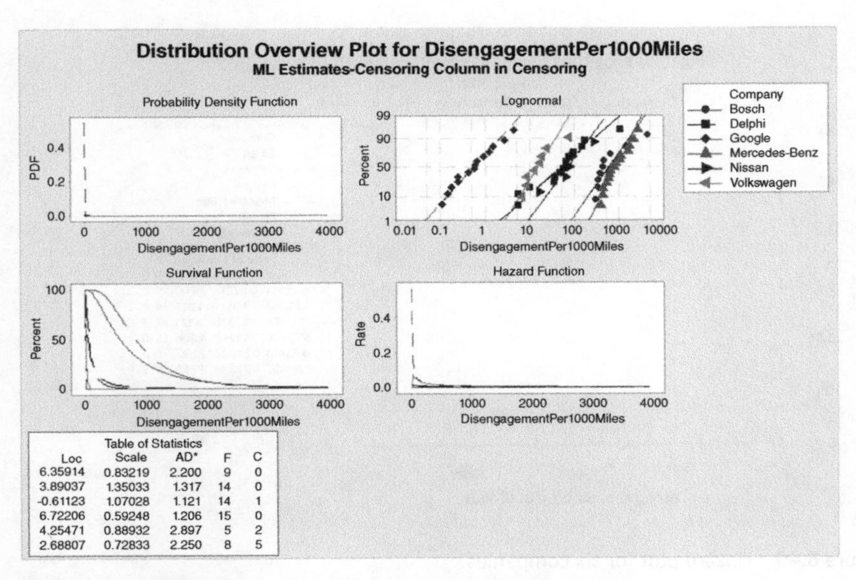

Table of Statistics

Loc	Scale	AD*	F	C
6.35914	0.83219	2.200	9	0
3.89037	1.35033	1.317	14	0
-0.61123	1.07028	1.121	14	1
6.72206	0.59248	1.206	15	0
4.25471	0.88932	2.897	5	2
2.68807	0.72833	2.250	8	5

Figure 8.38 Distribution overview plots for the six companies.

Since each chart's size is small, we can also produce an individual chart with a bigger scale. A single survival chart is created (Figure 8.39).

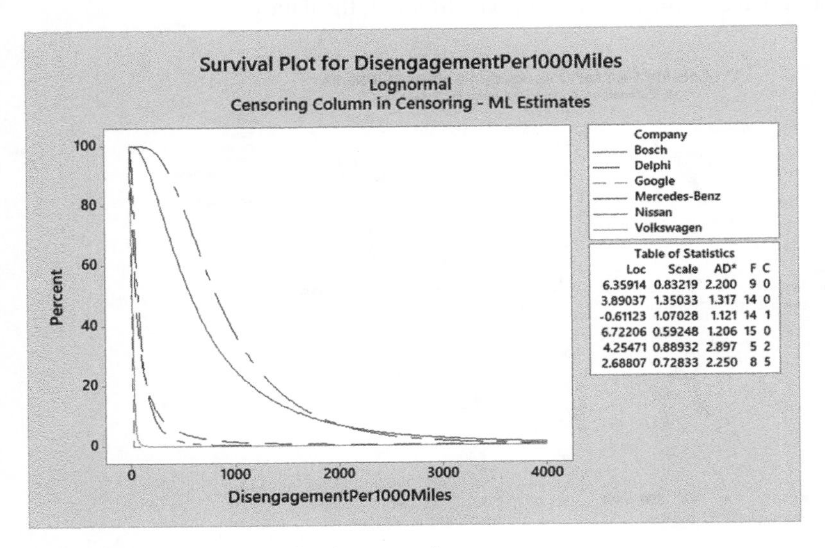

Figure 8.39 Survival plot for six companies.

The hazard plot can be constructed to compare the performance of the six companies (Figure 8.40).

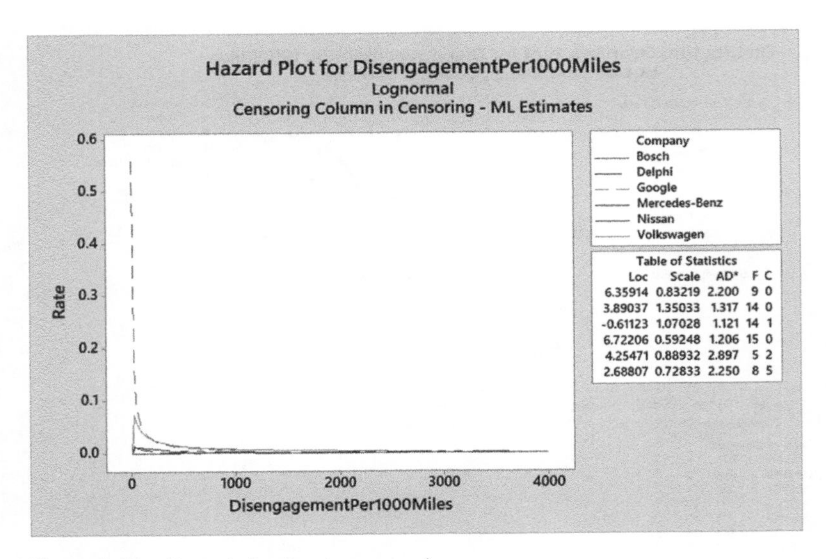

Figure 8.40 Hazard plot for six companies.

8.3.3 Python Practice

Python codes can be written to conduct the distribution ID plot. Figure 8.41 is an example of the codes for Google's disengagement data.

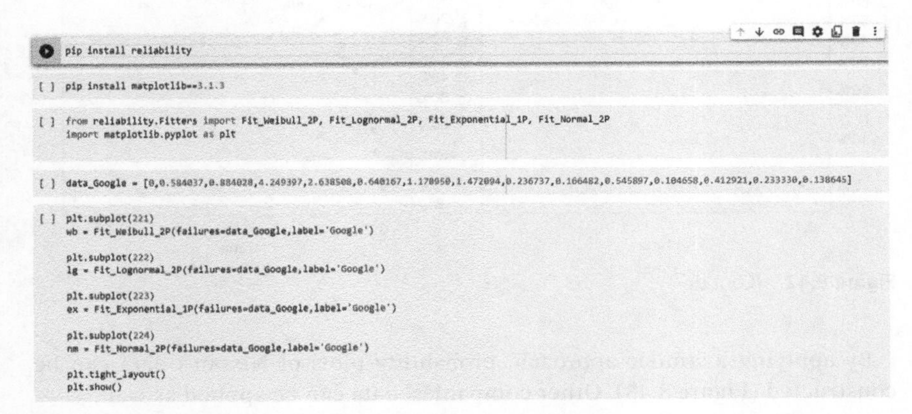

pip install reliability
[ ] pip install matplotlib==3.1.3
[ ] from reliability.Fitters import Fit_Weibull_2P, Fit_Lognormal_2P, Fit_Exponential_1P, Fit_Normal_2P
import matplotlib.pyplot as plt
[ ] data_Google = [0,0.584037,0.884028,4.249397,2.638508,0.640167,1.170950,1.472094,0.236737,0.166482,0.545897,0.104658,0.412921,0.233330,0.138645]
[ ] plt.subplot(221)
wb = Fit_Weibull_2P(failures=data_Google,label='Google')
plt.subplot(222)
lg = Fit_Lognormal_2P(failures=data_Google,label='Google')
plt.subplot(223)
ex = Fit_Exponential_1P(failures=data_Google,label='Google')
plt.subplot(224)
nm = Fit_Normal_2P(failures=data_Google,label='Google')
plt.tight_layout()
plt.show()

Figure 8.41 Python codes to run the distribution ID plot of Google's disengagement data.

After running all the codes, the probability plots with four different distributions can be created (Figure 8.42). Based on visual examination, the lognormal distribution could be selected.

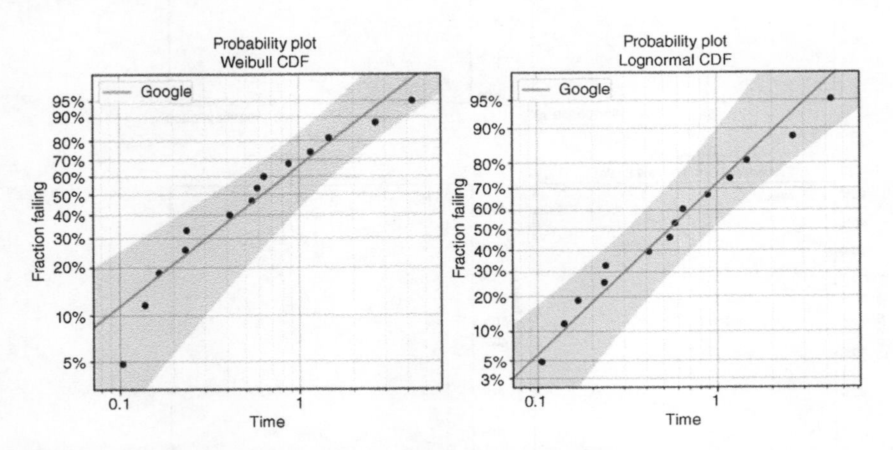

Figure 8.42 Probability plots of Google's disengagement data with Python.

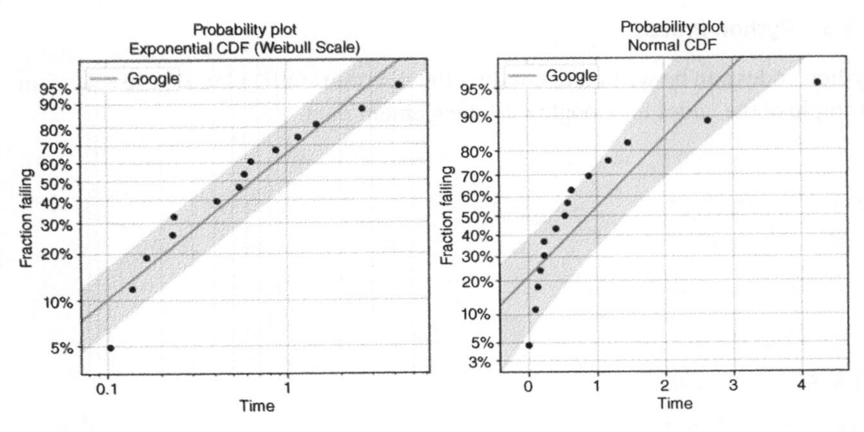

Figure 8.42 (Cont'd)

By applying a similar approach, probability plots of Nissan's data can be constructed (Figure 8.43). Other companies' data can be applied as well.

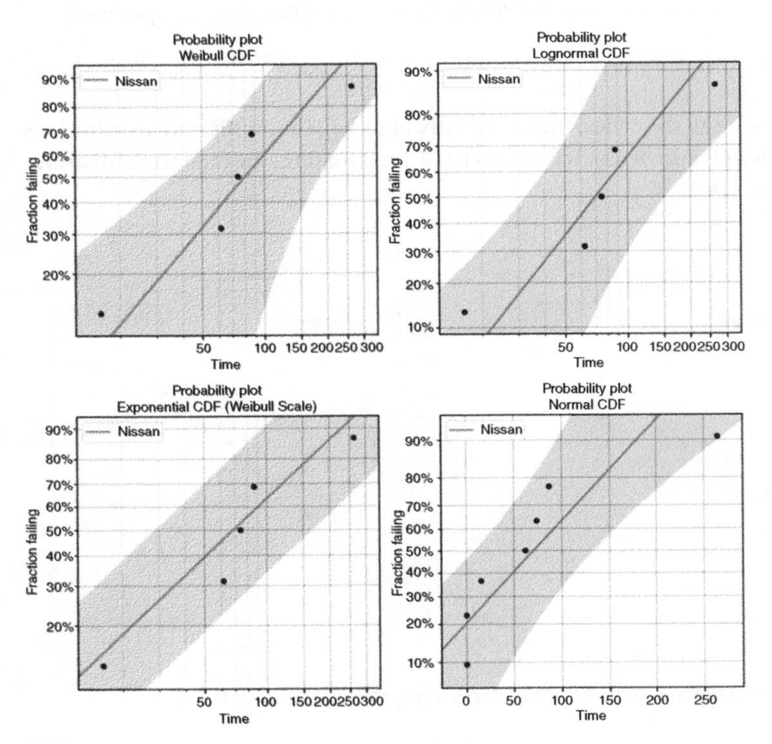

Figure 8.43 Probability plots of Nissan's disengagement data with Python.

With known parameters of the lognormal distribution of each company's data, the distribution overview charts can be constructed (Figure 8.44). Figure 8.44 shows the Python codes of Google's data.

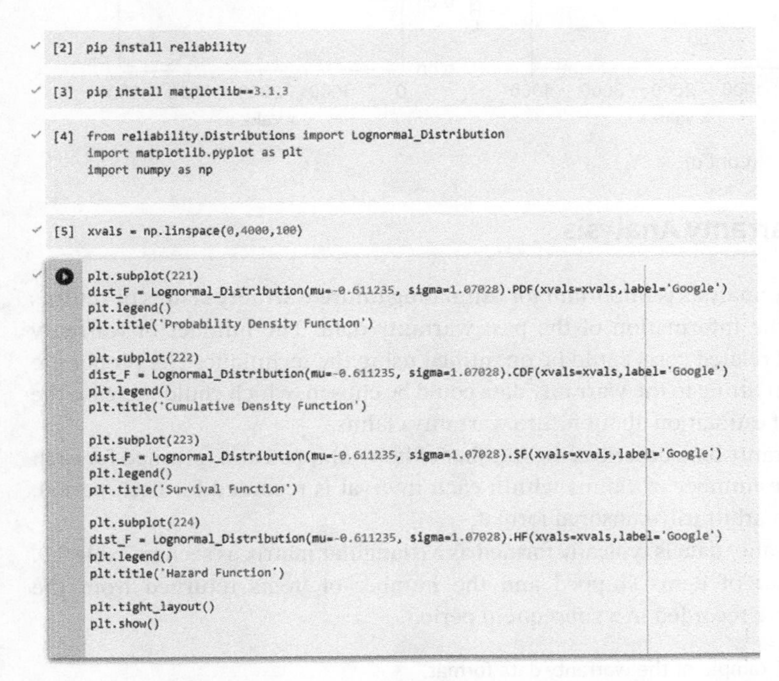

```
[2] pip install reliability

[3] pip install matplotlib==3.1.3

[4] from reliability.Distributions import Lognormal_Distribution
    import matplotlib.pyplot as plt
    import numpy as np

[5] xvals = np.linspace(0,4000,100)

    plt.subplot(221)
    dist_F = Lognormal_Distribution(mu=-0.611235, sigma=1.07028).PDF(xvals=xvals,label='Google')
    plt.legend()
    plt.title('Probability Density Function')

    plt.subplot(222)
    dist_F = Lognormal_Distribution(mu=-0.611235, sigma=1.07028).CDF(xvals=xvals,label='Google')
    plt.legend()
    plt.title('Cumulative Density Function')

    plt.subplot(223)
    dist_F = Lognormal_Distribution(mu=-0.611235, sigma=1.07028).SF(xvals=xvals,label='Google')
    plt.legend()
    plt.title('Survival Function')

    plt.subplot(224)
    dist_F = Lognormal_Distribution(mu=-0.611235, sigma=1.07028).HF(xvals=xvals,label='Google')
    plt.legend()
    plt.title('Hazard Function')

    plt.tight_layout()
    plt.show()
```

Figure 8.44 Python codes to conduct distribution overview plots of Google data.

After running all the codes, four different charts are created of Google data (Figure 8.45). A similar approach can be applied to each company's data.

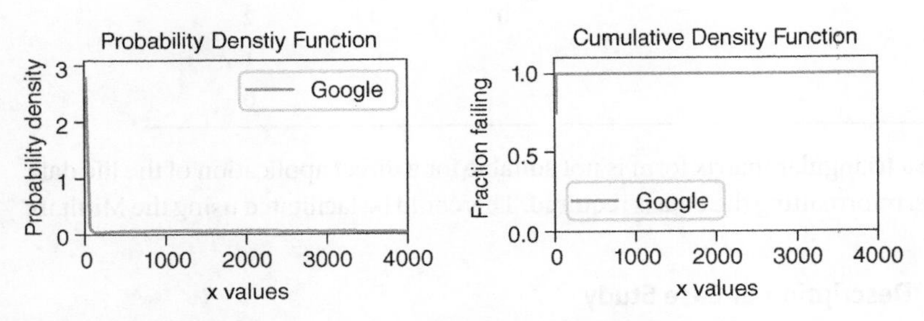

Figure 8.45 Distribution overview plots of Google's data with Python.

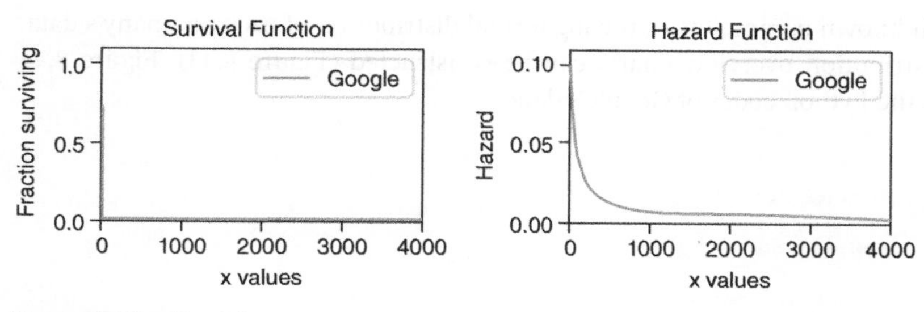

Figure 8.45 (Cont'd)

8.4 Warranty Analysis

A warranty analysis is important for estimating future warranty claims or returns based on the information of the past warranty data. The number of warranty claims and related costs could be quantified using the technique. The proper life distribution fitting to the warranty data could be chosen, which could increase the accuracy of estimation about future warranty claims.

The warranty data describes the number of items shipped and returned for each period. The number of claims within each interval is recorded for each period, which is an arbitrarily-censored format.

The warranty data is typically formed as a triangular matrix as seen in Table 8.9. The number of items shipped and the number of items returned from the shipment are recorded in a subsequent period.

Table 8.9 Example of the warranty data format.

Shipping Quantity	Month 1	Month 2	Month 3	Month 4	Month 5
1000	0	0	0	5	6
1000		0	0	0	3
1000			0	0	2
1000				0	1
1000					0

Since a triangular matrix form is not suitable for a direct application of the life data analysis, reformatting the data is required. This could be facilitated using the Minitab.

8.4.1 Description of Case Study

A reliability engineer in a vacuum company wants to estimate future warranty claims due to a defective part in a vacuum product. The engineer obtained warranty claim data for the past year. It is expected that future shipment of vacuum products per month will be 500. Table 8.10 shows historical warranty data in a triangular matrix.

Table 8.10 Historical warranty data of vacuum products.

Shipped	Month 1	Month 2	Month 3	Month 4	Month 5	Month 6	Month 7	Month 8	Month 9	Month 10	Month 11	Month 12
500	0	0	0	0	1	0	0	1	1	0	1	2
500		0	0	0	0	2	2	0	0	1	1	1
500			0	0	0	1	1	2	0	0	2	1
500				0	0	2	1	0	0	0	1	2
500					0	0	0	3	2	1	2	2
500						1	0	0	2	1	0	0
500							0	1	1	0	0	0
500								0	2	2	1	0
500									1	2	1	1
500										0	0	1
500											1	0
500												2

8.4.2 Minitab Practice

Step 1: Open Minitab. Prepare a data set (Figure 8.46).

↓	C1	C2	C3	C4	C5	C6	C7	C8	C9	C10	C11	C12	C13
	Shipped	Month 1	Month 2	Month 3	Month 4	Month 5	Month 6	Month 7	Month 8	Month 9	Month 10	Month 11	Month 12
1	500	0	0	0	0	1	0	0	1	1	0	1	2
2	500	*	0	0	0	0	2	2	0	0	1	1	1
3	500	*	*	0	0	0	1	1	2	0	0	2	1
4	500	*	*	*	0	0	2	1	0	0	0	1	2
5	500	*	*	*	*	0	0	0	3	2	1	2	2
6	500	*	*	*	*	*	1	0	0	2	1	0	0
7	500	*	*	*	*	*	*	0	1	1	0	0	0
8	500	*	*	*	*	*	*	*	0	2	2	1	0
9	500	*	*	*	*	*	*	*	*	1	2	1	1
10	500	*	*	*	*	*	*	*	*	*	0	0	1
11	500	*	*	*	*	*	*	*	*	*	*	1	0
12	500	*	*	*	*	*	*	*	*	*	*	*	2

Figure 8.46 Warranty data set in Minitab.

Step 2: Stat > Reliability/Survival > Warranty Analysis > Pre-Process Warranty Data (Figure 8.47).

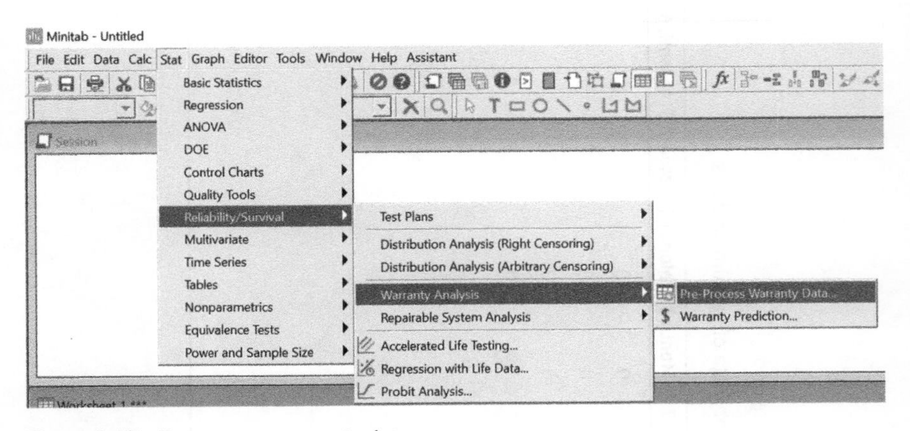

Figure 8.47 Pre-process warranty data.

Step 3: For the Data format, select Shipment values in a column (Figure 8.48). For the Shipment (sale) column, select Shipped. For the Return (failure) columns, select Month1–Month12.

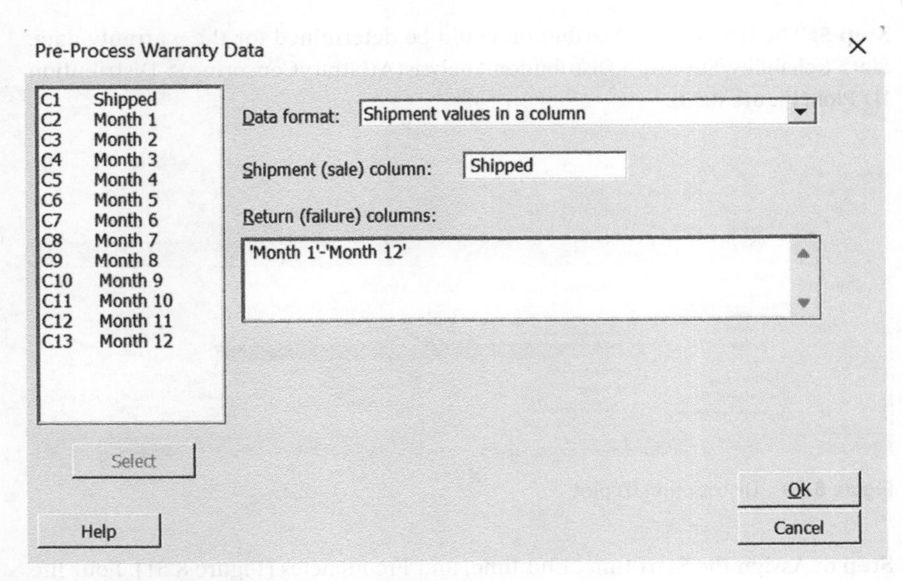

Figure 8.48 Pre-process warranty data setup.

Step 4: The reformatted data set (last three columns) is added (Figure 8.49). For example, a total of 5 vacuums were returned in the first month after the shipment.

↓	C1	C2	C3	C4	C5	C6	C7	C8	C9	C10	C11	C12	C13	C14	C15	C16
	Shipped	Month 1	Month 2	Month 3	Month 4	Month 5	Month 6	Month 7	Month 8	Month 9	Month 10	Month 11	Month 12	Start time	End time	Frequencies
1	500	0	0	0	0	1	0	0	1	1	0	1	2	0	1	5
2	500	*	0	0	0	0	2	2	0	0	1	1	1	1	2	5
3	500	*	*	0	0	0	1	1	2	0	0	2	1	2	3	7
4	500	*	*	*	0	0	2	1	0	0	0	1	2	3	4	9
5	500	*	*	*	*	0	0	0	3	2	1	2	2	4	5	7
6	500	*	*	*	*	*	1	0	0	2	1	0	0	5	6	5
7	500	*	*	*	*	*	*	0	1	1	0	0	0	6	7	2
8	500	*	*	*	*	*	*	*	0	2	2	1	0	7	8	4
9	500	*	*	*	*	*	*	*	*	1	2	1	1	8	9	6
10	500	*	*	*	*	*	*	*	*	*	0	0	1	9	10	2
11	500	*	*	*	*	*	*	*	*	*	*	1	0	10	11	2
12	500	*	*	*	*	*	*	*	*	*	*	*	2	11	12	2
13														1	*	498
14														2	*	499
15														3	*	499
16														4	*	495
17														5	*	495
18														6	*	498
19														7	*	496
20														8	*	490
21														9	*	494
22														10	*	493
23														11	*	493
24														12	*	494

Figure 8.49 Reformatted warranty data set.

Step 5: The best fitting distribution could be determined for the warranty data. Stat> Reliability/Survival> DistributionAnalysis(ArbitraryCensoring)> Distribution ID Plot (Figure 8.50).

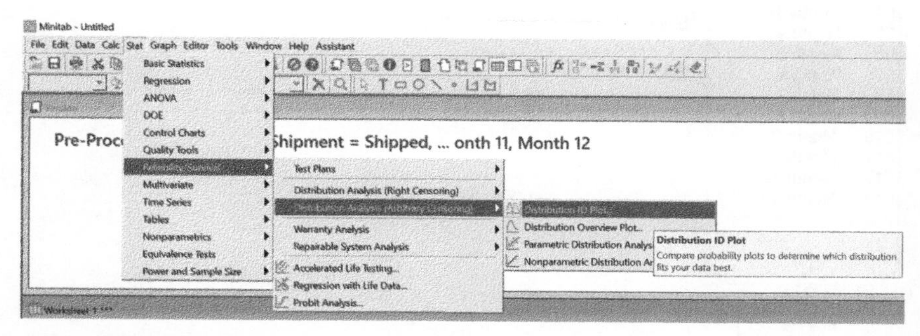

Figure 8.50 Distribution ID plot.

Step 6: Assign the Start time, End time, and Frequencies (Figure 8.51). Four life distributions could be applied for the fitness.

Figure 8.51 Distribution ID plot setup.

Step 7: Based on visual examinations of the probability plots, the Weibull distribution could be selected (Figure 8.52).

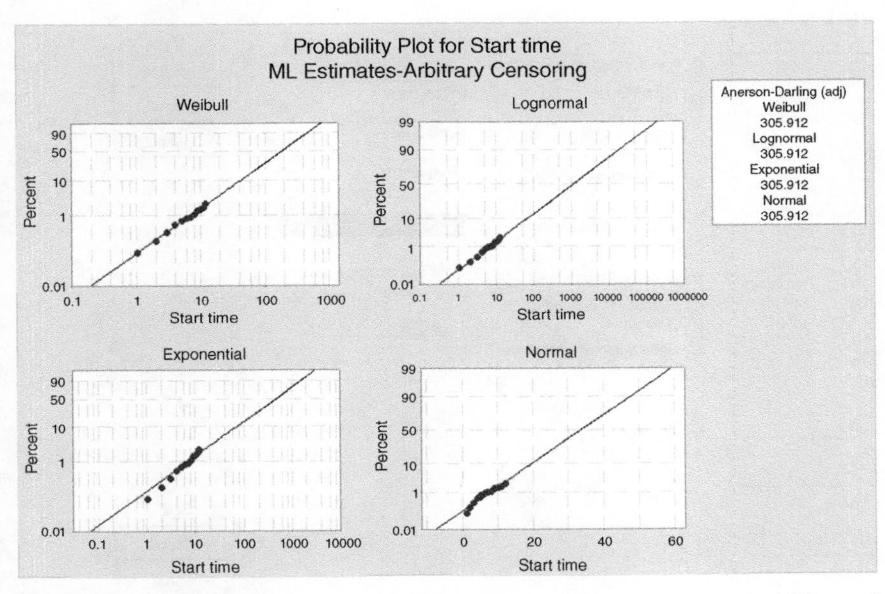

Figure 8.52 Probability plots.

Step 8: Stat > Reliability/Survival > Warranty Analysis > Warranty Prediction (Figure 8.53).

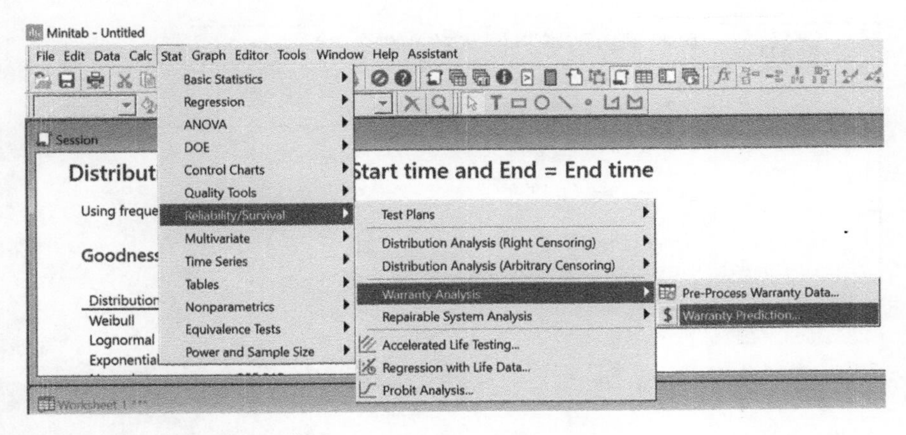

Figure 8.53 Warranty prediction.

Step 9: Start time, End time, and Frequencies are assigned (Figure 8.54). The Weibull distribution can be selected as an assumed distribution.

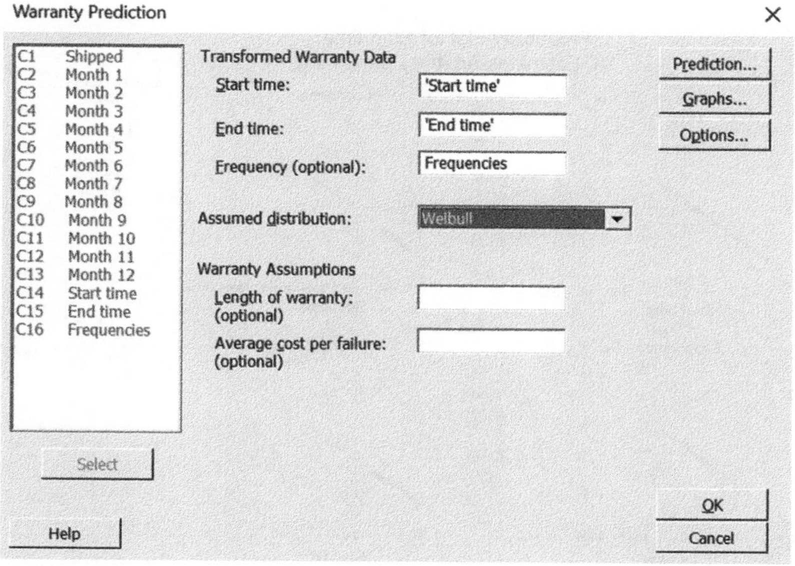

Warranty Prediction

C1	Shipped
C2	Month 1
C3	Month 2
C4	Month 3
C5	Month 4
C6	Month 5
C7	Month 6
C8	Month 7
C9	Month 8
C10	Month 9
C11	Month 10
C12	Month 11
C13	Month 12
C14	Start time
C15	End time
C16	Frequencies

Transformed Warranty Data
Start time: 'Start time'
End time: 'End time'
Frequency (optional): Frequencies
Assumed distribution: Weibull
Warranty Assumptions
Length of warranty: (optional)
Average cost per failure: (optional)
Prediction... Graphs... Options...
Select Help OK Cancel

Figure 8.54 Warranty prediction setup.

Step 10: Click the 'Prediction' tab (Figure 8.55). Set 500 as a production quantity for each time period. Hit OK.

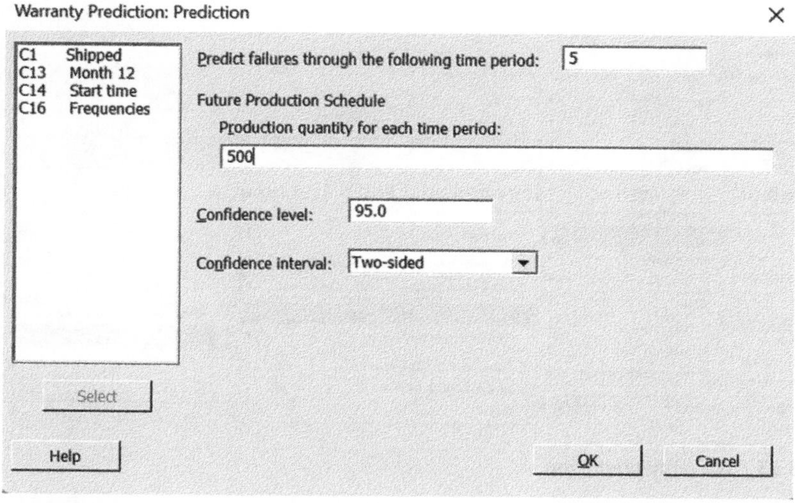

Warranty Prediction: Prediction

C1	Shipped
C13	Month 12
C14	Start time
C16	Frequencies

Predict failures through the following time period: 5
Future Production Schedule
Production quantity for each time period:
500
Confidence level: 95.0
Confidence interval: Two-sided
Select Help OK Cancel

Figure 8.55 Warranty prediction results.

Step 11: The summary of current warranty claims is shown in the session (Figure 8.56). The total number of units in the historical data was 6000, and 56 units failed. Based on the Weibull distribution estimation, 55 failures were estimated, which was close to the actual number of failures.

Summary of Current Warranty Claims

Total number of units	6000
Observed number of failures	56
Expected number of failures	55.3480
95% Poisson CI	(41.7356, 71.9837)
Number of units at risk for future time periods	5944

Figure 8.56 Summary of current warranty claims.

Step 12: Based on the table of the predicted number of failures, in five months from now, the expected number of warranty claims would be between 50 and 82 (Figure 8.57).

Table of Predicted Number of Failures

Future Time Period	Potential Number of Failures	Predicted Number of Failures	95% Poisson CI Lower	95% Poisson CI Upper
1	6444	10.7555	5.3197	19.3674
2	6944	22.5782	14.2451	34.0043
3	7444	35.4882	24.7862	49.2444
4	7944	49.5046	36.6845	65.3551
5	8444	64.6452	49.8541	82.4501

Figure 8.57 Table of predicted number of failures.

Step 13: The same information can be described using the predicted number of failures plot with 95% confidence interval (Figure 8.58).

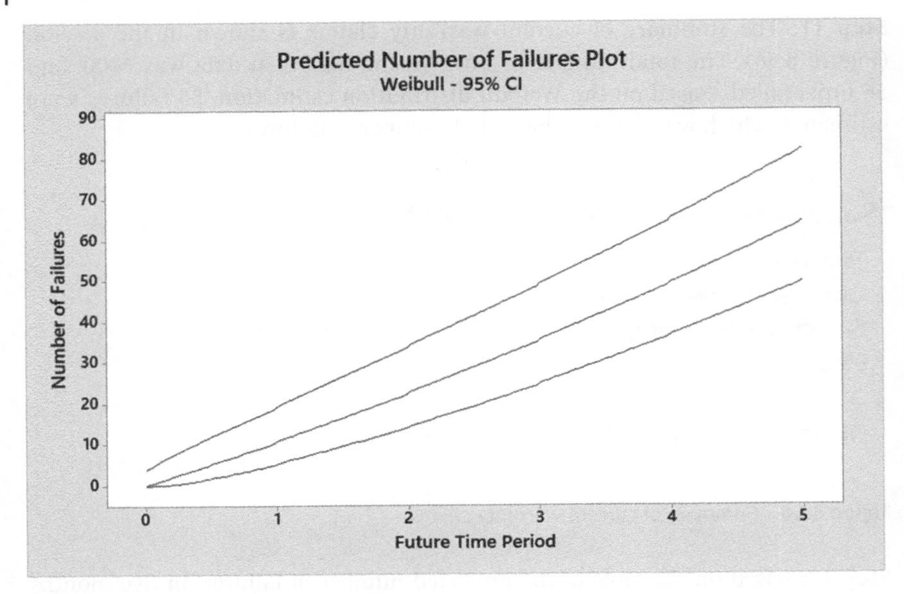

Figure 8.58 Predicted number of failures plot.

8.5 Stress–Strength Interference Analysis

In the Chapter 1, Introduction, it was mentioned that variation was one of the factors that could cause failure. This concept would apply to the stress–strength interference. If we know the specific probability distributions of each stress and strength, we could calculate the overlapped area at which the stress exceeds the strength.

If the normal distribution is a good fit for both stress and strength, the probability of failure can be calculated as:

$$\text{Probability of failure} = \Phi\left(\frac{\mu_{\text{strength}} - \mu_{\text{stress}}}{\sqrt{\sigma^2_{\text{strength}} + \sigma^2_{\text{stress}}}}\right) \tag{8.1}$$

8.5.1 Description of Case Study

A company obtained the strength data of 100 components. In addition, samples of the strength that were typically applied to the components were collected. Both strength and stress were well fitted by the normal distributions. For the stress data, $\mu = 50$, $\sigma = 20$. For the strength data, $\mu = 100$, $\sigma = 30$. Based on the variance and shape of the distribution in this case, calculate the probability of failure.

8.5.2 Minitab Practice

Go to Graph > Probability Distribution Plot (Figure 8.59).

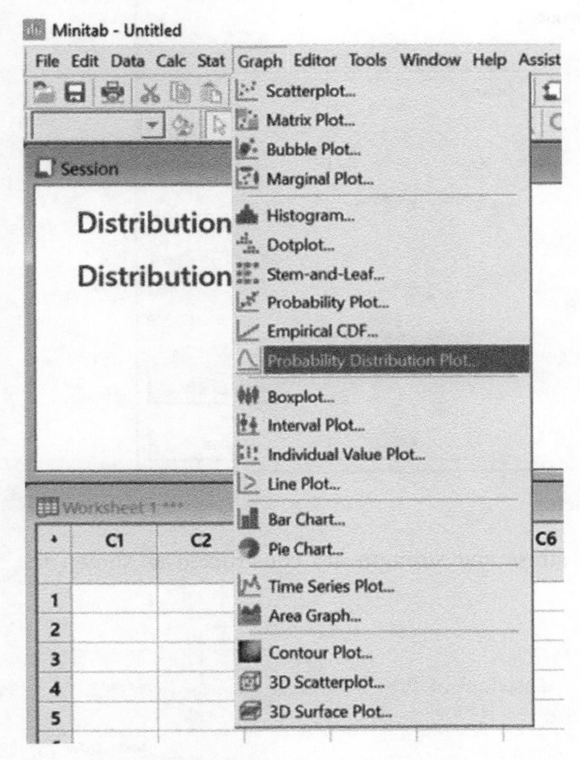

Figure 8.59 Probability Distribution Plot option under Graph.

Select 'Two Distributions' (Figure 8.60).

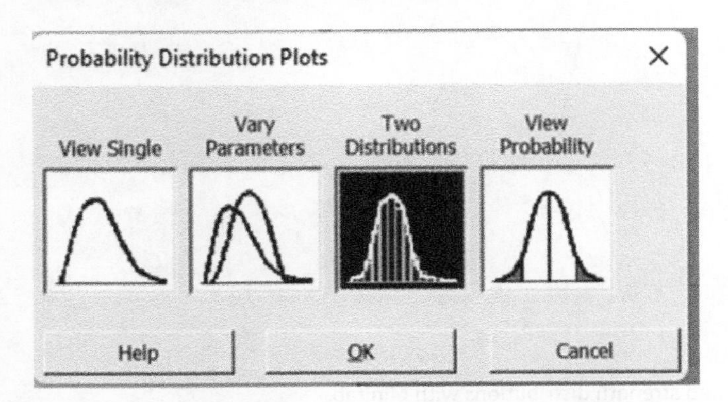

Figure 8.60 Two Distributions option under Probability Distribution Plots.

Select 'Normal' distribution for both Distributions 1 and 2 (Figure 8.61). Assign the mean and standard deviation of each distribution of stress and strength.

Probability Distribution Plot: Two Distributions

Distribution 1: Normal
Mean: 50
Standard deviation: 20

Distribution 2: Normal
Mean: 100
Standard deviation: 30

Multiple Graphs...
Help
OK
Cancel

Figure 8.61 Two distributions setup.

Two normal distributions of stress and strength are constructed as shown in Figure 8.62.

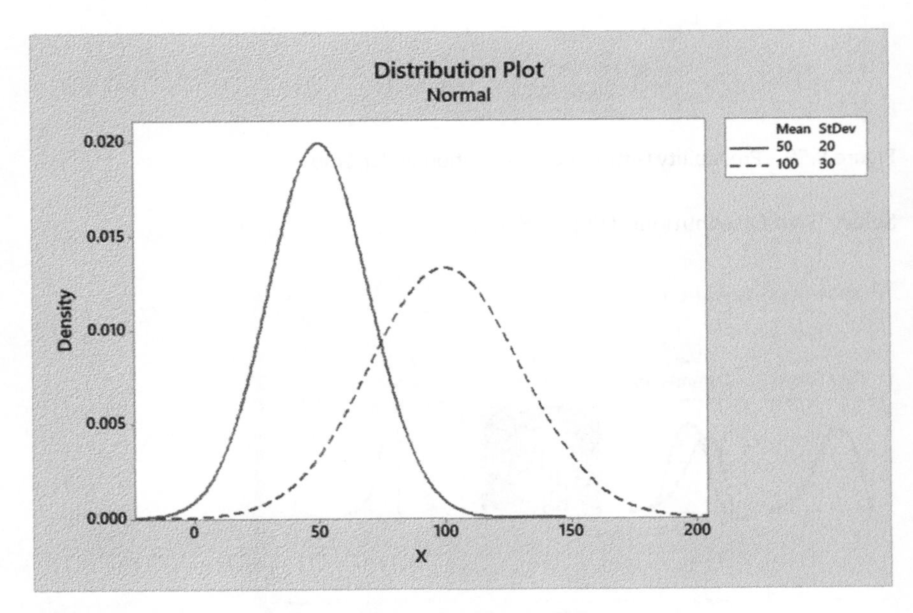

Figure 8.62 Stress and strength distributions with Minitab.

The probability of failure can be manually calculated.

$$\text{Probability of failure} = \Phi\left(\frac{\mu_{\text{strength}} - \mu_{\text{stress}}}{\sqrt{\sigma^2_{\text{strength}} + \sigma^2_{\text{stress}}}}\right) = \Phi\left(\frac{100-50}{\sqrt{30^2+20^2}}\right) = 0.082759 \quad (8.2)$$

8.5.3 Python Practice

Python can be used to construct two distributions and compute the probability of failure.

'stress_strength_normal' function can be imported (Figure 8.63).

'stress' and 'strength' could be set with the normal distribution with parameter values.

```
[ ] pip install reliability
```

```
[ ] pip install matplotlib==3.1.3
```

```
[ ] from reliability import Distributions
    from reliability.Other_functions import stress_strength_normal
    import matplotlib.pyplot as plt
```

```
stress = Distributions.Normal_Distribution(mu=50,sigma=20)
strength = Distributions.Normal_Distribution(mu=100,sigma=30)
stress_strength_normal(stress=stress, strength=strength)
plt.show()
```

Figure 8.63 Python codes used to construct the stress and strength distributions.

After running all the codes, PDFs of stress and strength are created, and the probability of failure is reported (Figure 8.64). The value is consistent with the previous manual calculation. There is an 8.28% chance of stress failure in the existing system.

Stress - Strength Interference

Stress Distribution: Normal Distribution (μ=50, σ=20)
Strength Distribution: Normal Distribution (μ=100, σ=30)
Probability of failure (stress > strength): 8.27589 %

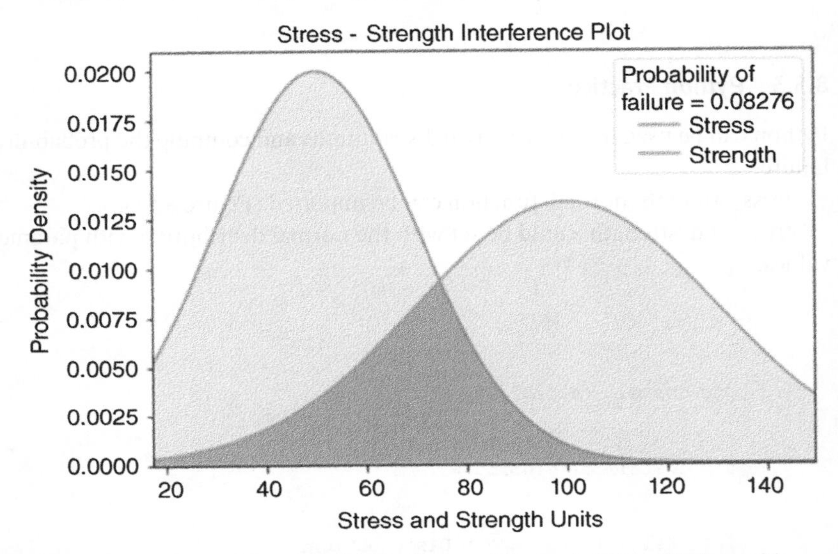

Figure 8.64 The stress and strength distributions and the probability of failure with Python.

8.6 Summary

- Parametric reliability analysis assumes that the reliability data follows certain types of statistical distributions.
- If none of the existing probability distributions fit the life data, nonparametric reliability analysis could be an alternative option. The Kaplan–Meier estimator is one of the commonly used methods in nonparametric reliability analysis.
- A warranty analysis is important for estimating future warranty claims or returns based on the information of the past warranty data. The number of warranty claims and related costs could be quantified using the technique.
- For the stress–strength interference, if we know the specific probability distributions of each stress and strength, we could calculate the overlapped area at which the stress exceeds the strength.

Exercises

1 An aerospace company collected the failure time of 20 randomly selected components as seen in the accompanying table. The testing was complete after 120 hours of observation. Three components were still operating at the end of testing. Determine the suitable distribution for the failure data. Estimate the time that 10% of the components would fail. Estimate the proportion of components that would survive after 50 hours. Apply both Minitab and Python.

Unit Number	Failure Time (hours)
1	41
2	46
3	54
4	56
5	60
6	62
7	64
8	75
9	77
10	86
11	89
12	91
13	95
14	98
15	102
16	103
17	106
18	Unknown
19	Unknown
20	Unknown

2 A textile manufacturing company collected warranty data of the previous year as seen in the accompanying table. The company's reliability engineer wants to estimate future warranty claims for the upcoming five months. A consistent shipping quantity of 500 textile units is expected for future production. Apply Minitab to find the future warranty claims.

Shipped	Month 1	Month 2	Month 3	Month 4	Month 5	Month 6	Month 7	Month 8	Month 9	Month 10	Month 11	Month 12
500	0	0	0	1	1	0	0	1	1	0	1	3
300		0	0	0	1	2	2	0	0	1	1	1
300			0	1	1	1	1	2	0	1	2	1
300				0	1	2	1	0	0	0	1	2
500					0	0	0	3	2	1	2	2
300						1	1	0	2	1	0	0
300							0	1	1	0	1	0
500								1	2	2	1	1
500									1	2	1	1
300										1	0	1
300											1	0
500												2

3 A company obtained the strength data of 500 components. Sample data of the strength that was typically applied to the components was collected. Both strength and stress were well fitted by the normal distributions. For the stress data, $\mu = 100$, $\sigma = 30$. For the strength data, $\mu = 200$, $\sigma = 50$. Based on the variance and shape of the distribution in this case, calculate the probability of failure.

4 For the description in Exercise 1, conduct a nonparametric reliability analysis using Minitab and Python. Compare the results with the parametric analysis results.

Reference

Nouri Qarahasanlou, A., Ataei, M., Khalokakaie, R., Ghodrati, B., and Jafarei, R. 2016. Tire demand planning based on reliability and operating environment. *International Journal of Mining and Geo-Engineering* 50(2):239–248.

Index

Reliability Analysis Using MINITAB and Python, First Edition. Jaejin Hwang.
© 2023 John Wiley & Sons, Inc. Published 2023 by John Wiley & Sons, Inc.
Companion Website: www.wiley.com\go\Hwang\ReliabilityAnalysisUsingMinitabandPython

Printed and bound by CPI Group (UK) Ltd, Croydon, CR0 4YY

16/04/2025

14658582-0002